Hydrological Extremes in a Changing Environment

In Focus – Special Book Series

Hydrological Extremes in a Changing Environment: Modelling and Attribution Analysis

Editors

Yanlai Zhou, Cosmo Ngongondo and Nils Roar Sælthun

IWA PUBLISHING

Published by

IWA Publishing
Unit 104–105, Export Building
1 Clove Crescent
London E14 2BA, UK
Telephone: +44 (0)20 7654 5500
Fax: +44 (0)20 7654 5555
Email: publications@iwap.co.uk
Web: www.iwapublishing.com

First published 2022

British Library Cataloguing in Publication Data
A CIP catalogue record for this book is available from the British Library

ISBN: 9781789063615

Contents

IWA PUBLISHING

 doi: 10.2166/nh.2022.101

Editorial: Hydrological extremes in a changing environment: modeling and attribution analysis

The International Symposium of Hydrological Sciences and High-efficiency Water Resources Utilization under the Changing Environment (ISHW 2019) was held in Wuhan, China, from October 24 to 26, 2019. It was co-organized by the School of Water Resources and Hydropower Engineering, and the State Key Laboratory of Water Resources and Hydropower Engineering Science, Wuhan University, and sponsored by the 111 Project 'Evolution Mechanism and Adaptive Strategies of Hydrological and Water Resources System under the Changing Environment' of the Ministry of Education of China and the Research Council of Norway (FRINATEK Project 274310).

The ISHW 2019 brought together over 300 participants from more than 30 universities, institutions, and enterprises around the world to address the issues related to hydrological sciences and water resources management under the changing environment. The participants had access to a multitude of diverse and exciting scientific sessions, social and networking events, and opportunities to interact with professional fellow scholars. In this symposium, a wide variety of cutting-edge topics centered around (1) Sponge City; (2) Water Resources Management; (3) Frequency Analysis of Nonstationary Hydrological Extremes; (4) Ecohydrology; (5) Climate Change; and (6) Information Technology (IT) and Artificial Intelligence (AI) in Hydrology and Water Resources.

This special issue publishes selected papers related to climate change and hydrological extremes, i.e. droughts and floods, which are globally important natural hazards with associated costly impacts on society and the environment. Floods and droughts result from the superposition of different processes at various space and time scales: physical processes in the atmosphere, catchments, the river systems, and anthropogenic activities. However, the characteristics of hydrological extremes have been altered due to climate change and variability, such that approaches for their detection, attribution, and the frequency of occurrence need to be revisited as they are no longer stationary processes. For more accurate estimation of hydrological extremes under nonstationary and uncertain conditions, there is a need for holistic assessments. Time–frequency analysis, hydrological modeling, physical-cause analysis, multivariate statistical analysis, and uncertainty analysis are powerful tools for detecting, attributing, and making frequency analysis of nonstationary hydrological extremes in a changing climate. Both nonstationarity and uncertainty of frequency analysis of extreme hydrological events should be integrated to reveal the possible operational alternatives to the assumption of stationarity in hydrological extremes frequency analysis.

This special issue of *Hydrological extremes in a changing environment: modeling and attribution analysis* includes eight papers, which are mainly grouped into two sections.

The first group of papers concentrates on hydrological drought issues under the nonstationary condition including the prediction of the standardized precipitation index (SPI) time series (Li *et al.* 2022a), investigation for the influencing factors of hydrological drought (Zhou & Zhou 2021), demonstration of the characteristics of propagation from the meteorological to hydrological drought (Ma *et al.* 2022), and the hydro-environmental response to the interbasin water resources development (Zhang *et al.* 2022). Li *et al.* (2022a) proposed a hybrid model coupled with singular spectrum analysis (SSA) and backpropagation artificial neural network (BP-ANN) named the SSA–BP-ANN model to overcome the limitation of the conventional ANN on predicting the nonstationary SPI time series. The proposed SSA–BP-ANN model is tested in case studies at three meteorological stations in Northern Shannxi Province, China. The results show that this model can produce more accurate predictions than the BP-ANN model, which has great potential for promoting drought early warning in arid regions. Zhou & Zhou (2021) investigated drought characteristics of the Pearl River basin during 1960–2015 including the drought duration, severity, intensity, affected area, and centroids based on the Standardized Evapotranspiration Deficit Index (SEDI) and three-dimensional clustering algorithm. The results reveal how these characteristics have affected net primary productivity (NPP) and improve the insight into the relationship between NPP and drought, which helps decision-makers manage droughts and provides guidance for drought-related studies across other regions. Ma *et al.* (2022) evaluated the characteristics and dynamics of drought propagation in different seasons and their linkages with key forcing factors through the SPI and the

Standardized Runoff Index (SRI). This study establishes the multiple linear regression model to quantitatively explore the influence of meteorological factors, underlying surface features, and teleconnection factors on the propagation time variations, and the Wei River Basin is selected as a study case. The results indicate that the differences of propagation time from meteorological to hydrological drought in different seasons and teleconnection factors including Arctic Oscillation, El Niño-Southern Oscillation, and Pacific Decadal Oscillation have strong influences on the propagation in autumn. Since interbasin water resource development has been exerting an increasingly evident impact on the hydro-environment of river basins, Zhang *et al.* (2022) selected the Han River as a study case to reveal the hydro-environmental response to China's interbasin water resources development. The results indicate that the runoff obviously decreases and low-flow duration significantly increases by 4–5 months after the operation of the South-to-North Water Transfer Project (SNWTP). Consequently, the flow decrease significantly contributed to the water quality deterioration in the middle and lower Han River.

The second group of papers concentrates on proposing novel approaches to the design flood analysis under the changing environment (Tan *et al.* 2021; Yan *et al.* 2021; Li *et al.* 2022b; Zhou *et al.* 2022). Yan *et al.* (2021) investigated the recommendation for the requirement on the length of data in nonstationary flood frequency analysis (NFFA). Twenty stations with record length longer than 100 years in Norway are selected and investigated by using both generalized extreme value (GEV)-ST and GEV-NS models with linearly varying location parameter (denoted by GEV-NS0). The results reveal that a minimum of 60 years of flood observations is recommended for NFFA for the selected basins in Norway. Zhou *et al.* (2022) proposed an integrated approach that combined the Generalized Additive Model for Location, Scale and Shape (GAMLSS) parametric method, the Copula function, and the Bayesian Uncertainty Processor (BUP) technique to make reliable probabilistic interval estimations of design floods, which is testified by the long-term (1954–2018) observed datasets of the Hanjiang River of China. The results demonstrate that the proposed approach can provide reliable probabilistic interval estimations of design floods, meanwhile reducing the uncertainty of NFFA. Tan *et al.* (2021) developed a nonlinear reservoir model in which the relationships of reservoir water level with reservoir volume and discharge are assumed to be nonlinear in order to accurately describe the routing process as it takes into consideration the interactions between hydrological loading and different discharge structures. The results show that the structure return period based on the linear model leads to an underestimation of the flood risk under the conditions of the high reservoir water level. Li *et al.* (2022b) presented the evaluation of a multisite statistical weather generator (MulGETS: Multisite weather Generator of École de Technologie Supérieure) based on its simulation effect of precipitation in the Yangtze River Basin, which effectively generates spatially correlated sequences of precipitation with maintaining the spatial and temporal distribution characteristics. The results verify the rationality of the MulGETS in downscaling precipitation in the Yangtze River Basin on the spatial and temporal scales.

All the papers collectively reflect hydrological extreme issues under the changing environment. The Guest Editors would like to acknowledge the invaluable contributions of all the anonymous reviewers, and the consistently helpful guidance and support of Emma Gulseven at IWA Publishing and the Emeritus Editor-in-Chief, Professor Chong-Yu Xu.

Guest Editors
Yanlai Zhou
Professor, State Key Laboratory of Water Resources and Hydropower Engineering Science, Wuhan University, Wuhan 430072, China.

Cosmo Ngongondo
Professor, Department of Geography and Earth Sciences, University of Malawi, Chancellor College, P.O. Box 280, Zomba, Malawi.

Nils Roar Sælthun
Emeritus Professor, Department of Geosciences, University of Oslo, P.O. Box 1047 Blindern, N-0316 Oslo, Norway.

REFERENCES

Li, S., Xie, J., Yang, X., Xue, R. & Zhao, P. 2022a Improving BP artificial neural network model to predict the SPI in arid regions: a case study in Northern Shaanxi, China. *Hydrology Research* **53** (3), 419–440.

Li, Y., Zeng, X. M. & Guo, J. 2022b Evaluation of a multisite weather generator on precipitation simulation in the Yangtze river basin. *Hydrology Research* **53** (1), 206–220.

Ma, L., Huang, Q., Huang, S., Liu, D., Leng, G., Wang, L. & Li, P. 2022 Propagation dynamics and causes of hydrological drought in response to meteorological drought at seasonal timescales. *Hydrology Research* **53** (1), 193–205.

Tan, Q., Mao, Y., Wen, X., Jin, T., Ding, Z. & Wang, Z. 2021 Copula-based modeling of hydraulic structures using a nonlinear reservoir model. *Hydrology Research* **52** (6), 1577–1595.

Yan, L., Xiong, L., Ruan, G., Zhang, M. & Xu, C. Y. 2021 Design flood estimation with varying record lengths in Norway under stationarity and nonstationarity scenarios. *Hydrology Research* **52** (6), 1596–1614.

Zhang, J., Guo, L., Huang, T., Zhang, D., Deng, Z., Liu, L. & Yan, T. 2022 Hydro-environmental response to the inter-basin water resource development in the middle and lower Han River, China. *Hydrology Research* **53** (1), 141–155.

Zhou, Y. & Zhou, P. 2021 Decline in net primary productivity caused by severe droughts: evidence from the Pearl River basin in China. *Hydrology Research* **52** (6), 1559–1576.

Zhou, Y., Guo, S., Xu, C. Y., Xiong, L., Chen, H., Ngongondo, C. & Li, L. 2022 Probabilistic interval estimation of design floods under non-stationary conditions by an integrated approach. *Hydrology Research* **53** (2), 259–278.

doi: 10.2166/nh.2021.026

Design flood estimation with varying record lengths in Norway under stationarity and nonstationarity scenarios

Lei Yan[a,b], Lihua Xiong[a,*], Gusong Ruan[c], Mengjie Zhang[d] and Chong-Yu Xu[e]

[a] State Key Laboratory of Water Resources and Hydropower Engineering Science, Wuhan University, Wuhan 430072, China
[b] College of Water Conservancy and Hydropower, Hebei University of Engineering, Handan 056021, China
[c] Norwegian Water Resources and Energy Directorate (NVE), Oslo, Norway
[d] Department of Earth and Environmental Engineering, Columbia University, New York, NY, USA
[e] Department of Geosciences, University of Oslo, N-0315 Oslo, Norway
*Corresponding author. E-mail: xionglh@whu.edu.cn

ABSTRACT

In traditional flood frequency analysis, a minimum of 30 observations is required to guarantee the accuracy of design results with an allowable uncertainty, however, there has not been a recommendation for the requirement on the length of data in NFFA (nonstationary flood frequency analysis). Therefore, this study has been carried out with three aims: (i) to evaluate the predictive capabilities of nonstationary (NS) and stationary (ST) models with varying flood record lengths; (ii) to examine the impacts of flood record lengths on the NS and ST design floods and associated uncertainties; and (iii) to recommend the probable requirements of flood record length in NFFA. To achieve these objectives, 20 stations with record length longer than 100 years in Norway were selected and investigated by using both GEV (generalized extreme value)-ST and GEV-NS models with linearly varying location parameter (denoted by GEV-NS0). The results indicate that the fitting quality and predictive capabilities of GEV-NS0 outperform those of GEV-ST models when record length is approximately larger than 60 years for most stations, and the stability of the GEV-ST and GEV-NS0 is improved as record lengths increase. Therefore, a minimum of 60 years of flood observations is recommended for NFFA for the selected basins in Norway.

Key words: average design life level, flood record length, nonstationary flood frequency analysis, predictive capabilities, split-sample testing

HIGHLIGHTS

- This study evaluates the predictive capabilities of nonstationary (NS) and stationary (ST) models with varying flood record lengths.
- This study examines the impacts of flood record lengths on the NS and ST design floods and associated uncertainties.
- A minimum of 60 years of flood observations is recommended for NFFA for the selected basins in Norway.

1. INTRODUCTION

Flooding is one of the major natural disasters and has caused casualties and huge economic losses worldwide in the 21st century (Hallegatte *et al.* 2013; Arnell & Gosling 2016; Kan *et al.* 2016; Slater & Villarini 2016; Kuang *et al.* 2018; Yin *et al.* 2018; Zhang *et al.* 2018; Yan *et al.* 2019a; Hou *et al.* 2020; Li *et al.* 2020; Venvik *et al.* 2020). In the hydrological communities, flood frequency analysis (FFA) is the most widely used approach to characterize flood risk and estimate design values for hydrological structures. The fundamental assumption of FFA is that the historical extreme flood events are independent and identically distributed, namely the so-called stationary (ST) assumption. However, it should be mentioned that the ST assumption is quite rigorous because of the dynamic nature of flood events caused by climate change and human activities (Vogel *et al.* 2011; Luke *et al.* 2017; Serago & Vogel 2018; Zhang *et al.* 2019). Thus, the ST assumption has been questioned by many researchers under changing environment (Milly *et al.* 2008; Villarini *et al.* 2009; Montanari *et al.* 2013; Milly *et al.* 2015; Yan *et al.* 2017a; Xu *et al.* 2018; Xiong *et al.* 2019; Yan *et al.* 2019b).

Nonstationary FFA (NFFA) is proposed to address the nonstationary (NS) issue of extreme flood events. As evidenced from recent review articles, there are several different methods for NFFA (Khaliq *et al.* 2006; Hall *et al.* 2013; Madsen *et al.* 2014; Bayazit 2015; Hao & Singh 2016; Yan *et al.* 2017b; Salas *et al.* 2018; Jiang *et al.* 2019; Qu *et al.* 2020; Xiong *et al.* 2020; Yan *et al.* 2021), among which the time-varying moments (TVM) method is the most popular approach. In the TVM method, the statistical parameters of flood distribution are no longer invariant but change with time or other covariates to capture the trend of flood frequency (Rigby & Stasinopoulos 2005). However, despite the numerous studies in NFFA, there is neither a noncontroversial method for performing NFFA nor well accepted governmental guidelines for updating current design flood strategies (Serago & Vogel 2018). One of the concerns is the tremendous uncertainties in NFFA induced by model structure and parameter estimation of NS models (Hu *et al.* 2015; Serinaldi & Kilsby 2015; Luke *et al.* 2017; Serago & Vogel 2018; Sen *et al.* 2020). Typically, the model structure of the NS model is more complex than that of ST model, since additional parameters are introduced to capture the historical trend of extreme flood events. Thus, theoretically, more flood samples are required to provide reliable and robust parameter estimations for NS models. However, in many cases, the available flood record length is shorter than 50 years and is often limited to 20–30 years (McCuen & Galloway 2010; Kjeldsen *et al.* 2014; Kobierska *et al.* 2017; Ekeu-Wei 2018; Hu *et al.* 2020). So, for practical reasons, it is necessary to investigate the impacts of record lengths on NFFA considering the various availability of flood observations worldwide.

Another concern is the predictive capabilities of NS models. Numerous literatures have reported that NS models can improve the statistical representation of historical flood series compared with ST models (Sun *et al.* 2018; Xie *et al.* 2018; Lu *et al.* 2019; Su & Chen 2019), but the predictive capabilities of NS models for out-of-sample prediction are still questioned. Only very limited study has compared the predictive capabilities of NS and ST models. Luke *et al.* (2017) conducted split-sample testing and comprehensively evaluated predictive capabilities of NS log-Pearson Type III distributions compared with ST models in the USA, and recommended the use of updated ST model for flood risk analysis in basins that have been affected by human activities. But they did not consider the effects of different flood record lengths on the predictive capabilities of NS models. In Norway, Kobierska *et al.* (2017) evaluated the impacts of record lengths on the performance of ST models with different combinations of distributions and estimation methods and found that there is no clear threshold of record length about using a three-parameter distribution or a two-parameter distribution under ST condition.

The last concern is the engineering design problem based on the NS model. Theoretically, we can predict future annual flood distribution by extrapolating the statistical parameters of the established NS model. As a result, the predicted future flood distributions vary from one year to another year, which brings about new challenge for design flood estimation. Under the framework of NFFA, if we still follow the design method under stationarity, we can obtain annual design flood for a given return period, which is impractical for engineering design since the relationship between design flood and return period is no longer one-to-one. To address the problem of design flood estimation in NFFA, researchers have carried out many studies and developed several design flood methods in recent years (Olsen *et al.* 1998; Parey *et al.* 2007, 2010; Cooley 2013; Rootzén & Katz 2013; Salas & Obeysekera 2014; Acero *et al.* 2017, 2018; Hu *et al.* 2018; Wang *et al.* 2019; Byun & Hamlet 2020; Lu *et al.* 2020; Yan *et al.* 2020). Yan *et al.* (2019a) comprehensively compared different methods and recommended the use of average design life level (ADLL) and equivalent reliability (Hu *et al.* 2018). Byun & Hamlet (2020) developed an NS Monte Carlo method to address NS design problems by explicitly accounting for the projected changes of extreme hydrometeorological events. However, the effects of flood record length on the estimation of NS design flood have not been reported, which deserves more attention considering the larger uncertainties in the calculation of design flood with longer return period using relatively short flood series (Kjeldsen *et al.* 2014; Machado *et al.* 2015; Engeland *et al.* 2017; Strupczewski *et al.* 2017; Hu *et al.* 2020; Kochanek *et al.* 2020).

The aforementioned concerns are all related to the record length of flood series. There is a consensus that all the sampling schemes, parameter estimation and design flood estimation can be significantly influenced by the available flood record length. In Norway, dams should be able to withstand extreme floods with high return periods, such as 500 years or 1,000 years, based on the dam safety regulations of Norway (Lovdata 2010). Besides, infrastructures should withstand 20-, 200-, or 1,000-year return period flood events, resting with the impacts of flood events (TEK10 2016). Under ST condition, typically, a minimum of 30 years of flood samples is required for FFA (Stedinger *et al.* 1993; Midttømme *et al.* 2011; Kobierska *et al.* 2017). Luke *et al.* (2017) conducted NFFA using 1,250 annual maximum discharge records throughout the USA with record length ranging from 30 to 65 years, in which the record length of about 75% series is between 35 and 45 years. Yan *et al.* (2017a) conducted NFFA using annual maximum flood series (AMFS) with record lengths of 59, 62 and 66. Parey *et al.* (2010) constructed the NS model using 106 years of annual maximum temperature data. However, when we turn to nonstationarity, what are the record length requirements for NFFA with the expected accuracy or uncertainty? And how the record length might influence the predictive capabilities of NS models and the estimation of NS design flood? In this study, we attempted to answer these questions. Therefore, the objectives of this study are: (i) to evaluate the predictive capabilities of NS and ST models with varying flood record lengths, (ii) to examine the effects of varying flood record lengths on the

NS and ST design floods and associated uncertainties and (iii) to provide guidance for the probable requirements of record length for NFFA with the expected accuracy or uncertainty. To achieve these objectives, 20 AMFS with a record length of larger than 100 years in Norway were selected for illustration purposes.

2. STUDY AREA AND DATA

In this study, a number of annual maximum flood records of hydrological stations in Norway was used for the purpose of evaluating the effects of varying record lengths on model performance, predictive capabilities and design flood estimation of GEV (generalized extreme value)-ST and GEV-NS0. Initially, we collected flood data of 74 hydrological stations throughout the Norwegian mainland from the Norwegian Water Resources and Energy Directorate (NVE). Figure 1 presents the geographical location and data information of the collected stations. Then, only the data sets with more than 100 observations were reserved to ensure the availability of at least 70 observations for the maximum record length in the fitting period and 30 observations in the evaluation period. Finally, the long flood records of 20 watersheds throughout Norway were chosen for illustration purposes (including 2 stations with 99 observations).

Hydrological variables are affected by both climate and land-use/land-cover changes to facilitate the study of climate change and its impact on hydrological variability, and a Hydrological Reference Dataset (HRD) consists of good-quality long-term data series not influenced by human activities causing non-climate-related variability or change was constructed by Norwegian Water Resources and Energy Directorate (NVE) (Fleig *et al.* 2013). HRD consists of 250 reference stations throughout Norway. One of the selection criteria is the degree of river basin development. Particularly, the proportion of

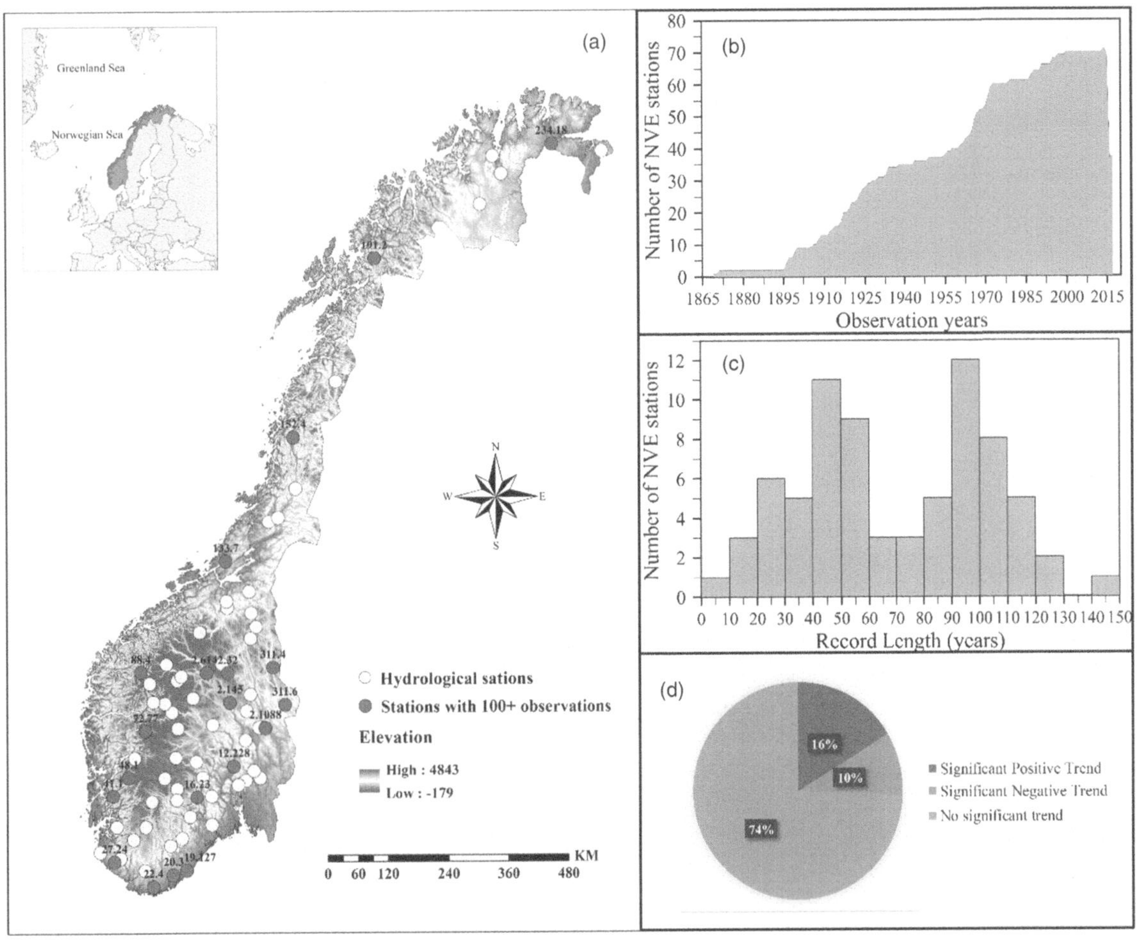

Figure 1 | (a) Location information of the hydrological stations and station with record length of larger than 100 years. (b) The number of available stations with observations in given years. (c) Histogram of record lengths of available stations in this study. (d) Trend detection results of available stations.

urban area of the 250 reference stations is lower than 10%. The other criterion is the absence of significant regulations, diversions, or water use. The selected 250 potential reference stations are classified as 'unregulated' in the reports on NVE's network of streamflow stations (Pettersson 2003; Kleivane 2006). In this study, two of the final selected nine stations are included in the HRD. For the remaining stations, similar criteria of no significant regulations and land-use/land-cover changes were used. For these reasons, the insensitive land-use and land-cover change as well as other minor physical changes, if any, are not explicitly considered in the study.

The main characteristics of the selected 20 watersheds, including the drainage areas, the geographical locations and starting and end years of the observations, are presented in Table 1. Previous studies have documented the existence of significant trend in the annual maxima and peaks over threshold in some regions of Norway (Vormoor *et al.* 2015, 2016; Debele *et al.* 2017). In this study, the trend-free pre-whitening Mann-Kendall test (TFPW-MK) (Yue *et al.* 2002) was employed to examine the trend of the selected AMFS by removing the effects of trend and serial correlation on the examination results (Table 1). The trend is statistically significant at 5% significant level if the *P*-value in Table 1 is smaller than 0.05. It should be mentioned that TFPW-MK suffered from type-1 error and Serinaldi & Kilsby (2016) proposed a corrected version called the TFPWcu method. See Serinaldi & Kilsby (2016) for the mathematical and numerical demonstration of TFPW flaws and the TFPWcu method, and see Serinaldi *et al.* (2020) for detailed discussion about the errors resulting from this superficial application of statistical analysis.

3. METHODOLOGY

In this section, we describe how the effects of varying record lengths are considered in NFFA from model construction to design flood estimation. The flowchart of this study is presented in Figure 2.

Table 1 | Data Information of the selected 20 watersheds with long record lengths in Norway

Station_ID	Name	Area (km²)	Data period	Record length	Longitude	Latitude	*P*-value
2.32	Atnasjø	463.3	1916–2015	100	10.22	61.85	0.095
2.145	Losna	11,200	1896–2017	122	10.28	61.33	0.022*
2.614	Rosten	1,833	1917–2015	99	9.40	61.86	0.685
2.1088	Glomma ved Skogmuseet	15,500	1871–2017	147	11.56	60.87	0.000*
12.228	Kistefoss	3,703	1917–2015	99	10.36	60.22	0.050*
16.23	Kirkevollbru	3,845.4	1906–2015	110	9.04	59.69	0.001*
19.127	Rygenetotal	3,946.4	1900–2015	116	8.67	58.41	0.011*
20.3	Flaksvatn	1,780	1900–2017	118	8.20	58.33	0.719
22.4	Kjæøemo	1,757.7	1897–2015	119	7.53	58.12	0.115
27.24	Helleland	184.7	1897–2015	119	6.15	58.53	0.009*
41.1	Stordalsvatn	130.7	1913–2015	103	6.01	59.68	0.340
48.1	Sandvenvatn	470	1908–2017	110	6.55	60.01	0.502
72.77	Flåm bru	263	1908–2017	110	7.12	60.84	0.020*
88.4	Lovatn	234.9	1900–2015	116	6.89	61.86	0.003*
133.7	Krinsvatn	206.6	1916–2015	100	10.23	63.80	0.187
152.4	Fustvatn	525.7	1909–2015	107	13.31	65.91	0.019*
191.2	Øvrevatn	526	1914–2015	102	17.94	68.86	0.679
234.18	Polmak	14,161.4	1912–2015	104	28.02	70.07	0.161
311.4	Femundsenden (Femunden)	1,790	1896–2017	122	11.94	61.92	0.081
311.6	Nybergsund	4,424.9	1909–2015	107	12.32	61.26	0.413

'*' denotes the trend is significant at 5% significance level.

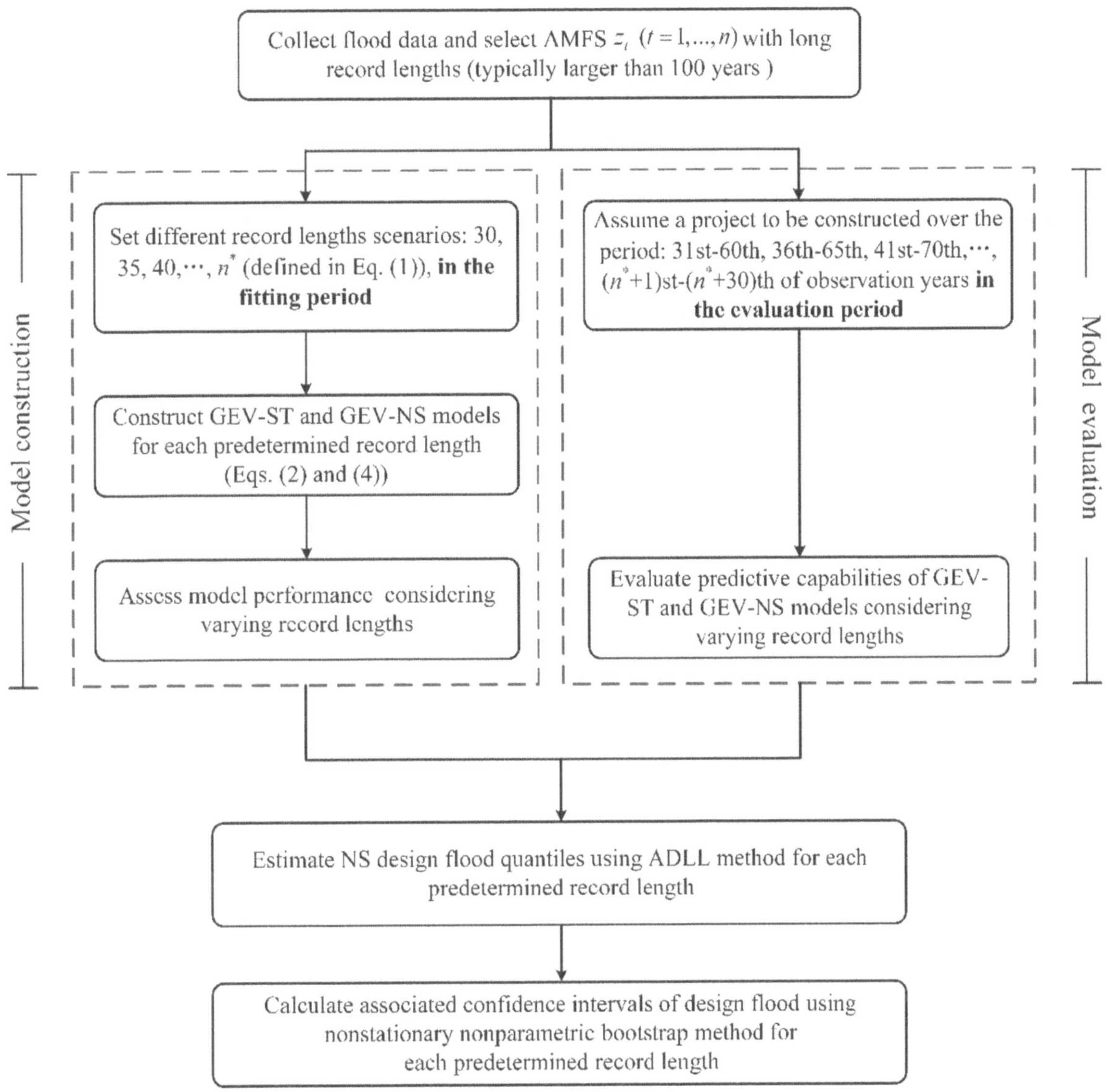

Figure 2 | The flowchart of this study.

3.1. Setting of varying record length scenarios and the split-sample testing method

To assess the effects of record lengths on model performance, predictive capabilities and design flood estimation, different record length scenarios are generated. For an observed AMFS $z_t(t = 1, \ldots, n)$, we assume there is a hydrological project to be constructed with a design life of 30 years within the observation years. Thus, the predetermined observations $z_t^f(t = 1, \ldots, n^*)$ with record length n^* in the fitting period and assumed future floods $z_t^e(t = n^* + 1, \ldots, n^* + 30)$ with a record length of 30 in the evaluation period are determined from the subset of the entire AMFS. The predetermined record lengths n^* are sampled every 5 years from 30 to $n^*_{\max}$, which is given by the following equation:

$$n^*_{\max} = 5 \times \text{floor}[(n - 30)/5] \tag{1}$$

where function floor returns the round down of $(n - 30)/5$, and 30 is the design life of a project. Herein the split-sample testing method is used. For example, if there are 106 observations, $n^*_{\max}$ is equal to 75 and thus the record lengths are set to be 30, 35, 40, …, 75. Subsequently, both ST GEV (GEV-ST) and NS GEV (GEV-NS) models are built for each predetermined record length in the fitting period, and the following 30 years' observations are assumed to be the design lifespan of a hydrological project, which is retained for evaluating the predictive capabilities and estimating the design flood in the evaluation period (Figure 3).

3.2. NFFA with varying record lengths (fitting period)

3.2.1. NS GEV model

In this study, we focus on the NFFA using single distribution rather than the nonstationarity caused by abrupt change or mixed flood generation mechanisms. GEV distribution is often used to model the annual maxima, such as AMFS, under the ST context. In the fitting period, for the AMFS $z_t^f(t = 1, \ldots, n^*)$, the cumulative distribution function (CDF) of GEV-ST model is given by the following equation:

$$\begin{cases} G(z_t^f|\mu, \sigma, \varepsilon) = \exp\{-[1 + \varepsilon(z_t^f - \mu)/\sigma]^{-1/\varepsilon}\} \\ 1 + \varepsilon(z_t^f - \mu)/\sigma > 0 \end{cases}, t = 1, \ldots, n^* \tag{2}$$

where μ, $\sigma > 0$ and ε are the location, scale and shape parameters of GEV-ST model, respectively. However, under NS condition, the statistical parameters of GEV model can be modeled as a function of covariates, such as time, to capture the trend of AMFS. The CDF of GEV-NS model is as follows:

$$\begin{cases} G_t(z_t^f|\mu_t, \sigma_t, \varepsilon_t) = \exp\{-[1 + \varepsilon_t(z_t^f - \mu_t)/\sigma_t]^{-1/\varepsilon_t}\} \\ 1 + \varepsilon_t(z_t^f - \mu_t)/\sigma_t > 0 \end{cases}, t = 1, \ldots, n^* \tag{3}$$

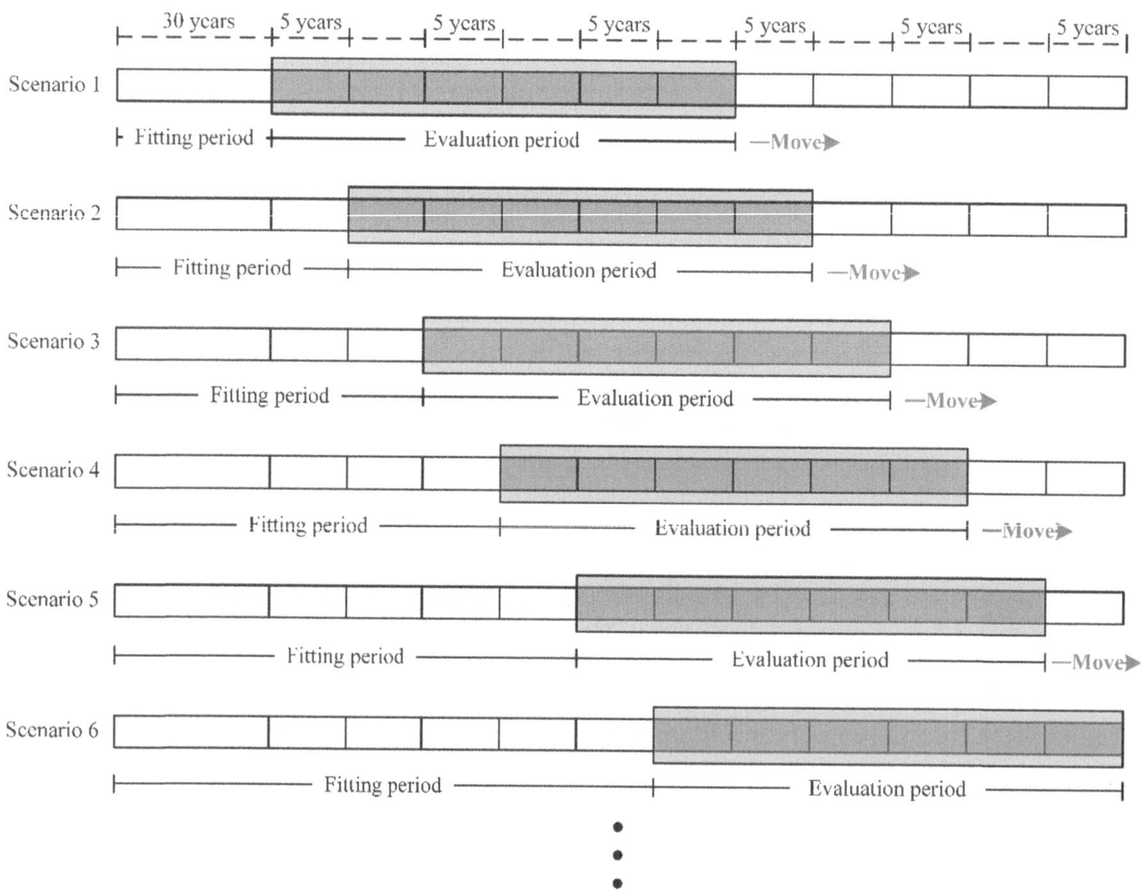

Figure 3 | Schematic diagram of varying record length scenarios using moving windows. The yellow rectangular boxes indicate the predetermined record lengths in the fitting period, increased by 5 years in each scenario. The light blue boxes represent the assumed 30-year lifespan of a project, moving with the increase of predetermined record lengths. It should be noted that there are more scenarios in our study than the six scenarios displayed above. Please refer to the online version of this paper to see this figure in color: http://dx.doi.org/10.2166/nh.2021.026.

where $G_t(\cdot)$ denotes the time-varying CDF of GEV-NS model. μ_t, $\sigma_t > 0$ and ε_t are the time-varying location, scale and shape parameters of GEV model, respectively, and t is the time scale. The time-varying statistical parameters, varying from one year to another year, are estimated based on the whole trend of the entire AMFS. Theoretically, although all three statistical parameters can be time-varying, the shape parameter in Equation (3) is sensitive and difficult to estimate (Cheng & Aghakouchak 2014; Du *et al.* 2015). Besides, the aim of this study is not to select the optimal model for a single basin, but to investigate the impacts of record lengths in NS frequency analysis. Thus, the structure of GEV-NS models should be simplified and fixed for each record length. The GEV-NS models only considering the time-varying location parameter become one of the most widely used GEV-NS models in practical applications. This kind of simplified GEV-NS models can generate realistic design quantiles consistent with the probabilistic behavior of extreme events (Cheng & Aghakouchak 2014; Sarhadi & Soulis 2017), which is described as follows:

$$G_t^{-1}(1-p|\mu_t, \sigma, \varepsilon) = \mu_0 + \mu_1 \times t + \frac{\sigma}{\varepsilon} \times [(-\ln(1-p))^{-\varepsilon} - 1],\ t = 1, \ldots, n^* \tag{4}$$

where $G_t^{-1}(\cdot)$ is the time-varying inverse CDF of GEV-NS model. p is the annual exceedance probability. μ_0 and μ_1 are the model parameters to model the change of location parameter μ_t, that is $\mu_t = \mu_0 + \mu_1 \times t$. The GEV-NS model in Equation (4) is simplified and convenient, considering that the shape parameter ε of GEV is quite sensitive and difficult to be estimated. Therefore, in this study, the GEV-NS model is constructed based on Equation (4) as done by previous researchers (Zwiers *et al.* 2010; Westra *et al.* 2013; Luke *et al.* 2017; Um *et al.* 2017).

3.2.2. Assessment of model performance for the fitting period

In this study, both GEV-ST and GEV-NS models with linearly varying location parameter, denoted by GEV-NS0, are built for each predetermined record length. Thus, to assess the fitting qualities and compare the difference between GEV-ST and GEV-NS0, the Akaike information criterion (AIC) (Akaike 1974) is employed, which is described as follows:

$$\mathrm{AIC} = -2\ln(\zeta) + 2\rho \tag{5}$$

where ρ is the number of independently adjusted parameters of the model. ζ is the maximum likelihood value of the established model. It should be noted that both the statistical parameters of GEV-ST and GEV-NS0 are estimated by the maximum likelihood method. Thus, ζ can be estimated by calculating the maximum value of likelihood function via deriving the partial derivatives of the log-likelihood function, which is, for instance, $\ln L(\theta|z_t^f) = \ln \prod_{t=1}^{n^*} g(z_t^f|\mu, \sigma, \varepsilon)$ for the GEV-ST model.

3.3. Evaluation of predictive capabilities with varying record lengths (evaluation period)

For each predetermined record length, the likelihood function is used to evaluate the predictive capabilities of both GEV-ST and GEV-NS0. Theoretically, the likelihood function can provide information about how plausible the estimated statistical model is with the observations $z_t^e(t = n^* + 1, \ldots, n^* + 30)$ during the evaluation period. For the GEV-ST model, the likelihood function $L(\theta|z_t^e)$ for the evaluation period is described as follows:

$$L(\theta|z_t^e) = \prod_{t=n^*+1}^{n^*+30} g(z_t^e|\mu, \sigma, \varepsilon) \tag{6}$$

where $g(\cdot)$ is the probability density function (PDF) of GEV, and n^* is the record length of predetermined observations. However, the likelihood function $L(\theta_t|z_t^e)$ of GEV-NS model for the evaluation period is defined as follows:

$$L(\theta_t|z_t^e) = \prod_{t=n^*+1}^{n^*+30} g(z_t^e|\mu_t, \sigma, \varepsilon) \tag{7}$$

where θ_t represents the time-varying statistical parameters. It can be seen from Equations (6) and (7) that under ST condition, the likelihood function is evaluated based on a set of invariant statistical parameters, whereas under NS condition, the likelihood function is evaluated based on variant statistical parameters.

Besides, the probabilistic coverages were also used to assess the predictive capabilities of GEV-ST and GEV-NS0. Theoretically, the percentage of flood observations below the prescribed percentiles, namely 5th, 25th, 50th, 75th and 95th percentiles, should be 5, 25, 50, 75 and 95%. The absolute difference between theoretical and actual probabilistic coverages, denoted by D_{pc}, can be calculated by the following equation:

$$\begin{cases} D_{\mathrm{pc}} = |\mathrm{PC}_T - \mathrm{PC}_A| \\ \mathrm{PC}_A = \mathrm{countif}(z_t < \mathrm{percentiles})/n \end{cases} \tag{8}$$

where PC_T and PC_A are the theoretical and actual probability coverages, respectively, and $\mathrm{PC}_T = 5, 25, 50, 75$ and 95%. percentiles denote the 5th, 25th, 50th, 75th and 95th percentiles. n represents the number of observations retained in the evaluation period and equals 30 in this study.

3.4. NS design floods and uncertainties with varying record lengths

3.4.1. Average design life level

In this study, the ADLL method is used to solve the NS hydrological design problem using GEV-NS0. The ADLL method was proposed by Yan *et al.* (2017a) based on the concept of annual average reliability, denoted by $\mathrm{RE}^{\mathrm{ave}}_{T_1-T_2}$, under NS context. $\mathrm{RE}^{\mathrm{ave}}_{T_1-T_2}$ is defined as follows (Read & Vogel 2015):

$$\mathrm{RE}^{\mathrm{ave}}_{T_1-T_2} = \frac{1}{T_2 - T_1 + 1}\sum_{t=T_1}^{T_2}(1 - p_t) = \frac{1}{T_2 - T_1 + 1}\sum_{t=T_1}^{T_2} G_t(z_{\mathrm{q}}|\theta_t) \tag{9}$$

where p_t is the time-varying nonexceedance probability, z_{q} is the design quantile over the design life of a hydrological project to be constructed from T_1 to T_2. The ADLL method assumes that the annual average reliability corresponding to z_{q} should be equal to the yearly reliability $1 - 1/m$ for the return period m. Thus, the m-year design quantile $z^{\mathrm{ADLL}}_{T_1-T_2}(m)$ estimated by the ADLL method can be determined by the following equation:

$$\frac{1}{T_2 - T_1 + 1}\sum_{t=T_1}^{T_2} G_t(z^{\mathrm{ADLL}}_{T_1-T_2}(m)|\theta_t) = 1 - 1/m \tag{10}$$

ADLL is recommended for practical use because it has associated design floods with the design life period of projects and is able to yield reasonable design quantiles and confidence intervals (CIs). See Yan *et al.* (2017a), which compared different NS design methods and recommended the ADLL method, for details of the advantages of ADLL. It should be noted that the maximum record length is equal to the full record length when analyzing the effects of record lengths on design flood quantiles and associated CIs. In addition, the coefficient of variation (CV) of design flood is also estimated for each record length using the nonstationary nonparametric bootstrap (NNB) method to assess the stability of GEV-ST and GEV-NS0.

3.4.2. Uncertainty analysis using the NNB method

In this study, to comprehensively evaluate the effects of record lengths on the ST and NS design results, the uncertainties or CIs of design floods estimated by the ADLL method are also calculated based on the NNB method. The advantage of NNB model is that it depends strictly on the available data without any hypotheses and can be implemented regardless of model complexity. In the NNB method, first, an NS model is used to fit the original observations. Then, the original observations are transformed to identically distributed residuals, and the transformed residuals are resampled with replacement and back-transformed to obtain the new samples. Next, the same NS model constructed in the first step is employed to fit the bootstrapped new samples and estimate corresponding design quantiles. Repeat the above steps for a large number of times and finally determine the CIs of design quantiles. See Obeysekera & Salas (2014) and Yan *et al.* (2017a) for detailed information and operating steps about the NNB method.

4. RESULTS AND DISCUSSION

4.1. NFFA with varying flood record lengths

According to the results of the TFPW-MK test, 9 of the 20 selected AMFS exhibited a significant increasing or decreasing trend at 5% significant level, as displayed in Figure 4. Then, NS frequency analysis of the AMFS of the nine stations was

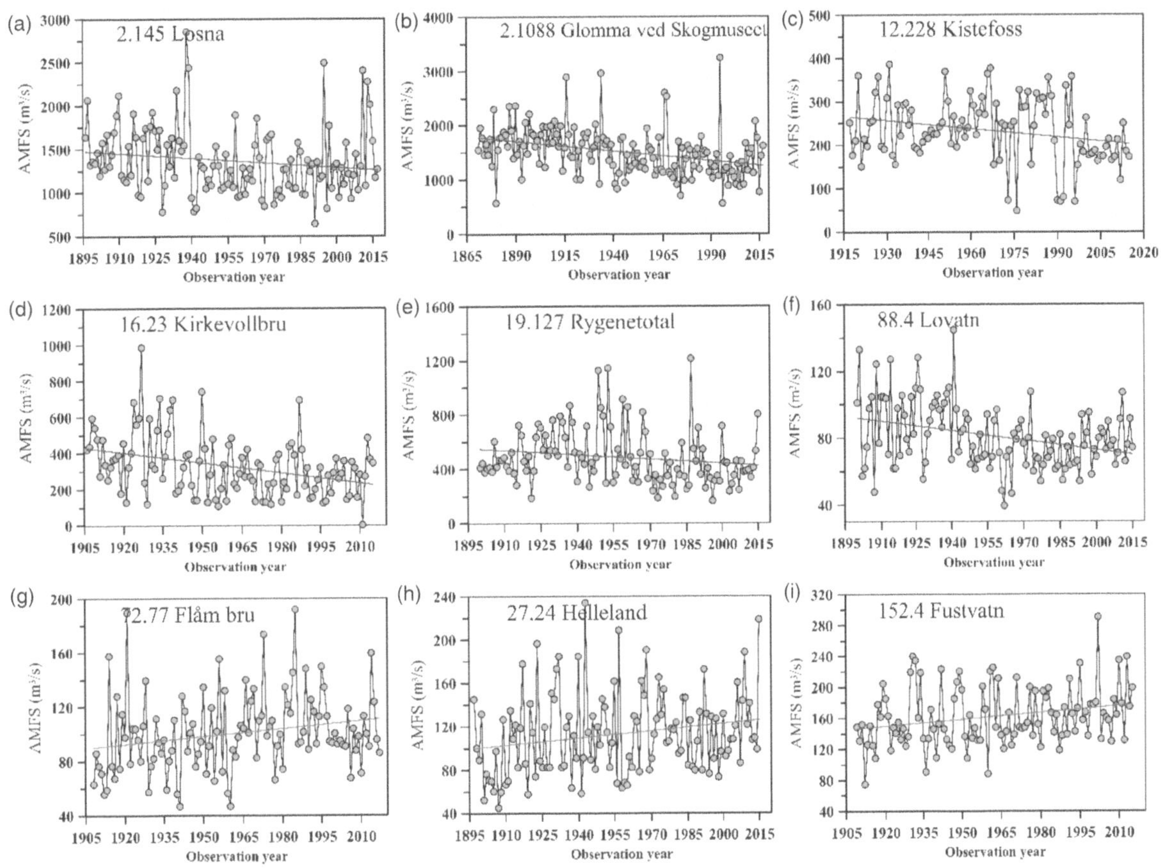

Figure 4 | AMFS of the selected nine stations with increasing or decreasing trends.

carried out using the generalized additive models for location, scale and shape (GAMLSS) package available in the R language platform.

Both GEV-ST and GEV-NS0 were constructed for different record length scenarios defined in Section 3.1. We evaluated and compared the model performance of GEV-ST and GEV-NS0 using the AIC values. As shown in Figure 5, for eight out of the nine stations, the AIC values of GEV-ST were eventually larger than those of GEV-NS0 with the increase of record lengths, indicating the better performance of GEV-NS0 particularly for samples with large record lengths. Furthermore, the differences between AIC values of GEV-ST and GEV-NS became larger with the increase of record lengths except for stations 12.228 and 19.127. Quantitatively, for 8 out of the 9 stations, the biggest AIC differences in Figure 5 are larger than 5, and for 2 out of the 9 stations, the biggest AIC differences are around or larger than 10, which confirms the advantage of the GEV-NS0. Besides, we also found that the superiority of GEV-NS0 starts to emerge roughly for record lengths of larger than 60 years, indicating the necessity of collecting more flood samples in NFFA rather than the minimum requirement of 30 samples in conventional FFA.

4.2. Predictive capabilities of GEV-ST and GEV-NS models with varying record lengths

Likelihood values were employed to evaluate the predictive capabilities of GEV-ST and GEV-NS0. As shown in Figure 6, for majority stations (six out of nine stations), the likelihood values of GEV-NS0 are eventually higher than those of the GEV-ST models for larger record length, particularly when the record length is larger than 60 years. It should be mentioned that the likelihood values in Figure 6 may be affected by high uncertainty for small samples. Besides, we also examined the difference between theoretical probability coverage and actual probability coverages which are the percentage of flood observations under 5th, 25th, 50th, 75th and 95th percentiles, denoted by D_5, D_{25}, D_{50}, D_{75} and D_{95}, respectively. As shown in Figures 7 and 8, the average value of D_5 is smaller than 10% for both GEV-ST and GEV-NS0 when record length is larger than 60, while the average value of D_{95} is smaller than 5% when record length is larger than 40, indicating the good predictive

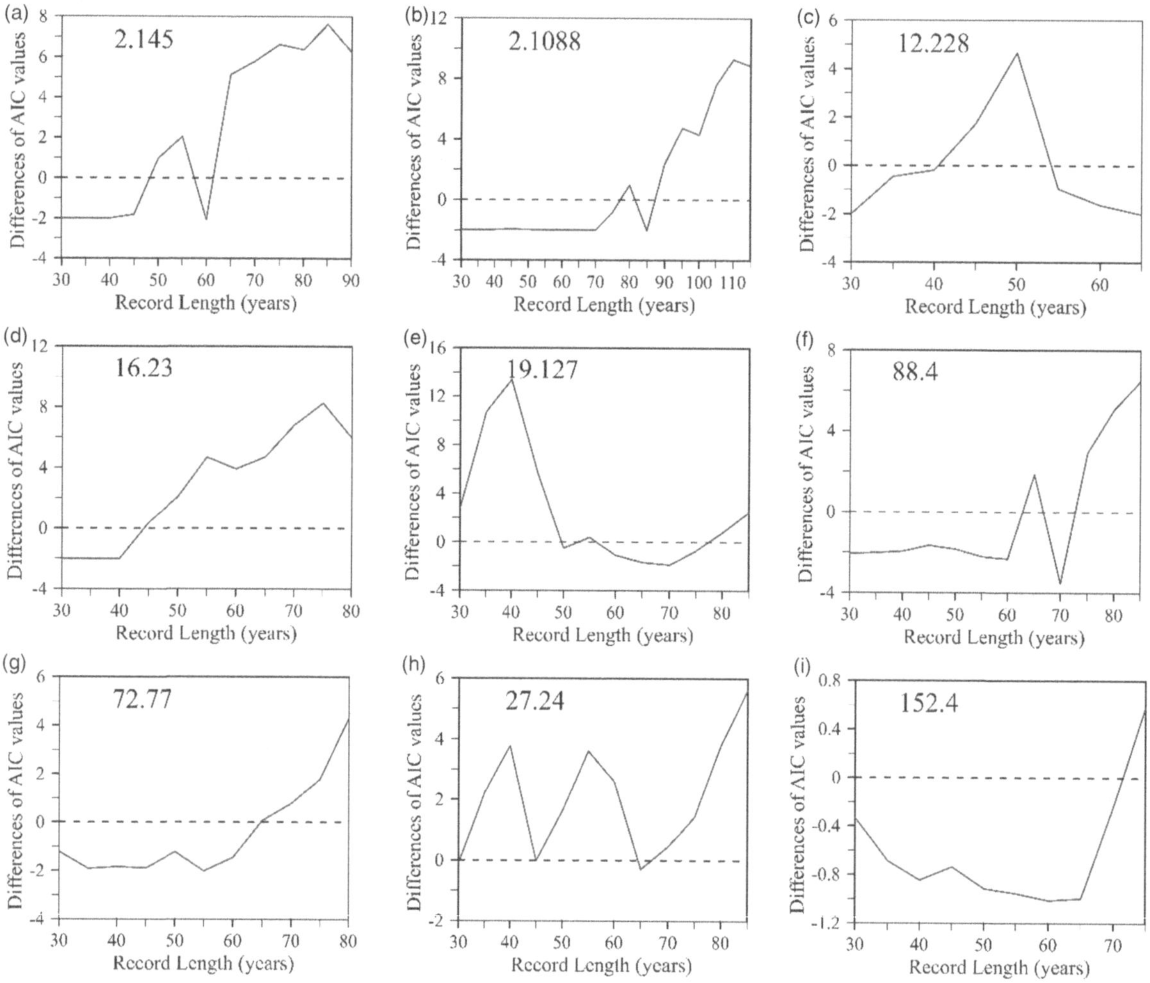

Figure 5 | Absolute differences of the AIC values calculated using the AIC values of GEV-ST models minus those of GEV-NS models. The values larger than zero indicate the better performance of GEV-NS models, and vice versa.

capabilities of extreme events for both GEV-ST and GEV-NS0. Furthermore, the results indicate that the difference in nearly all five plots tends to be smaller with the increase of record lengths for GEV-NS0, particularly when the record length is roughly longer than 60 years. While for the GEV-ST models, there is no significant decrease of the five plots with the increase of record lengths, and D_{25} and D_{50} even exhibit upward trends. It should be mentioned that the difference of probability coverage in Figure 7 may be affected by less stations for larger record lengths. Particularly, for D_{25} and D_{50}, the decreasing trends become more significant for the two largest sample sizes of 90 and 95. The '95-size' value relies on station 2.1088, while the '90-size' average is taken over stations 2.145 and 2.1088. In fact, decreasing trend can still be identified regardless of these two record lengths, particularly compared with the increasing trends of D_{25} and D_{50} for the GEV-ST model.

The above results confirm the better predictive capabilities of GEV-NS0 compared with GEV-ST models for most stations. The enhanced model performance and predictive capabilities of GEV-NS0 can be attributed to their better capture of the variability of extreme flood events when new flood samples are added into the flood population.

4.3. Influence of varying record lengths on NS design floods and uncertainties

After the evaluation of model performance and predictive capabilities of both GEV-ST and GEV-NS0, we further examine the influence of record length on design flood quantiles and associated CIs. Assuming the design life of hydrological structures is 30 years, both GEV-ST and GEV-NS0 were employed to estimate design floods using the ADLL method for each record length.

For each station, we first evaluated the stability of the established GEV-ST and GEV-NS0 by calculating the CVs of the 20-, 50- and 100-year design flood quantiles for each record length. Figure 9 shows the evolution of CV as a function of record

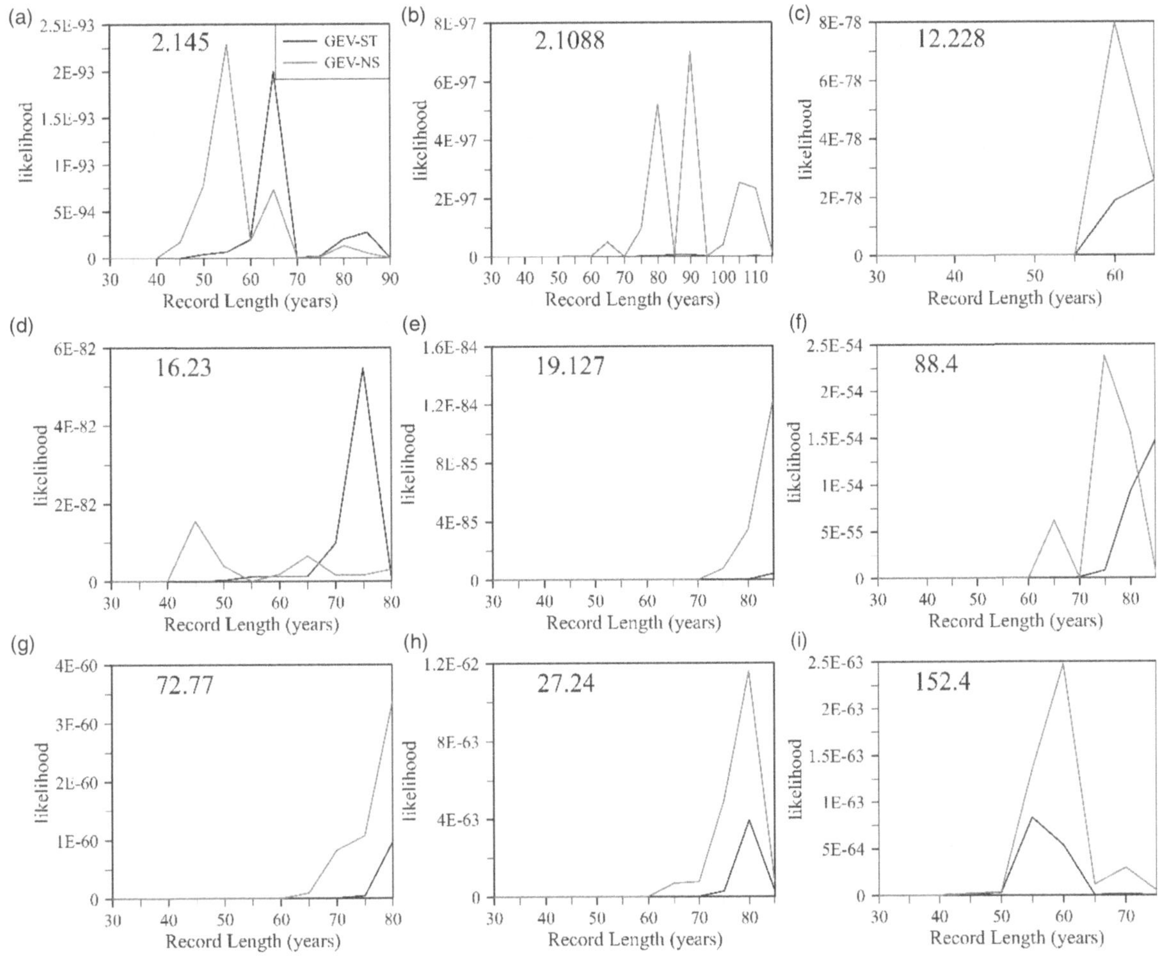

Figure 6 | Likelihood values of GEV-ST (black lines) and GEV-NS (red lines) models. Please refer to the online version of this paper to see this figure in color: http://dx.doi.org/10.2166/nh.2021.026.

length for all the selected nine stations. It is shown that for both ST and NS conditions, the CVs of 20-, 50- and 100-year design floods tend to be smaller with the increase of record length for nearly all stations, showing the increasing stability of the established statistical models as record lengths increase. Besides, when the record length is approximately larger than 50–60 years, the fluctuation of CV becomes gentle for most stations. It should be mentioned that for stations 19.127, 27.24 and 152.4, the stabilization of both models occurs for a record length of smaller than 60. As for the comparison of the stability of GEV-ST and GEV-NS0, the CVs of GEV-ST are smaller than those of GEV-NS for all the three return periods, which is due to the more complex model structure of GEV-NS0.

Figure 10 summarizes the evolution of 100-year design floods and associated 95% CIs with the increase of record length under both ST and NS contexts. The design values yielded by the GEV-ST model and the simplified GEV-NS model based on the ADLL method are similar, particularly with the increase of record length. For six out of the nine stations, the difference between 100-year design flood of ST and NS models is relatively small with the increase of record length or even becomes smaller when record length is larger than 50–60 years, see stations 12.228 and 19.127. Thus, it is difficult to discriminate between these two design strategies for a larger record length. For the rest of the stations, a representative example is station 2.1088, for which the difference between ST and NS design floods becomes larger when the record length is larger than 70 years and the largest difference is obtained when record length is 145 years. In terms of CIs, it is found that the associated bootstrapped 95% CIs of the 100-year design flood become smaller as the record lengths increase for both GEV-ST and GEV-NS0, showing the advantage of collecting more flood samples under both ST and NS conditions. In addition, there is no significant difference between the CIs of ST and NS design floods for most stations.

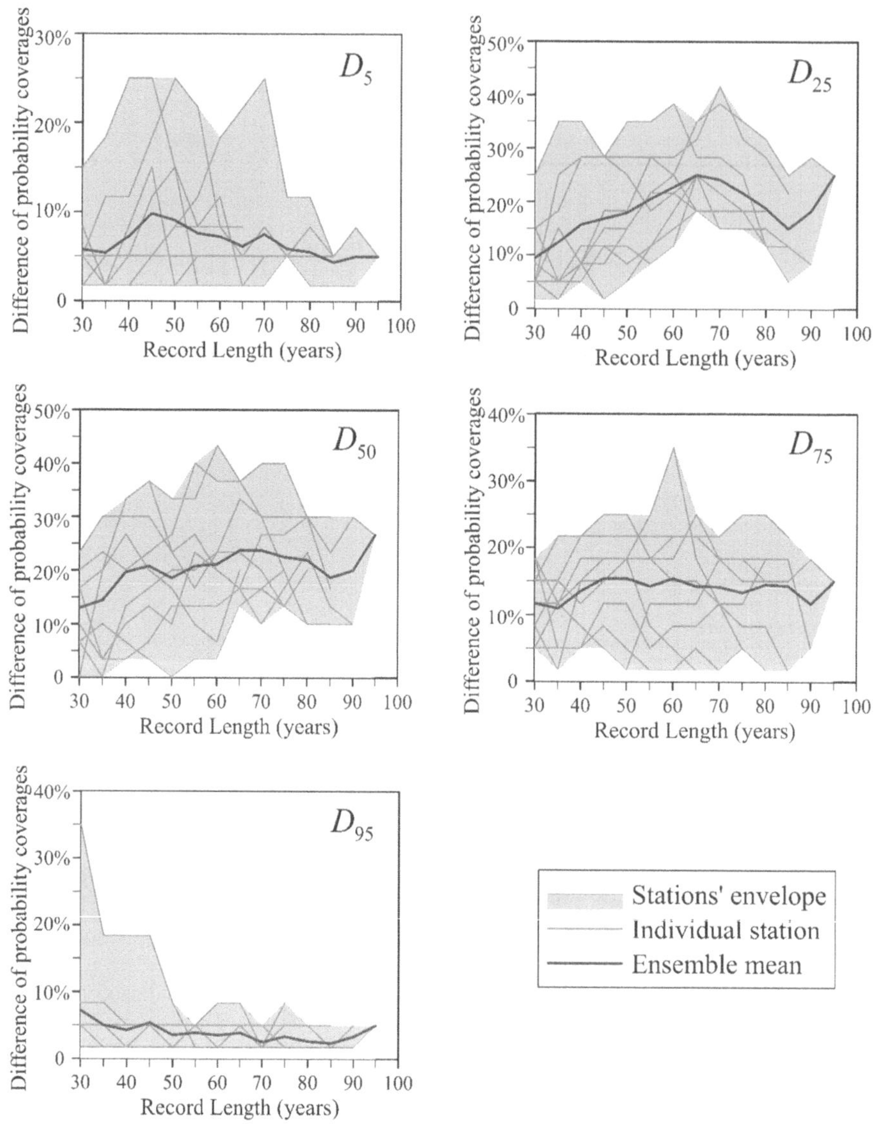

Figure 7 | The difference between theoretical probability coverage and actual probability coverages and the percentage of flood observations under 5th, 25th, 50th, 75th and 95th percentiles, denoted by D_5, D_{25}, D_{50}, D_{75} and D_{95}, respectively, using GEV-ST models. The difference was calculated using Equation (8), and ensemble mean was the average difference of nine stations.

4.4. Discussions

There is still a hot academic debate on whether we should switch to NS frequency analysis, considering the complex model structure, prediction of future distributions and the uncertainties associated with NS hydrological design. However, there is no doubt that we are living in a changing world. Thus, this study focuses on what we could do if the world is NS. Again, it should be emphasized that nonstationarity cannot be inferred only by trend tests but should be confirmed by attribution analysis, statistical analysis and empirical analysis (Milly *et al.* 2015; Yan *et al.* 2017a). As discussed by Serinaldi *et al.* (2018), trend analysis can at most be used as statistical tools for preliminary screening whose outcome should be carefully checked and complemented with exogenous information. If a clear physical mechanism related to a predictable evolution of the properties of the process at hand is not identified, we cannot make conclusions about the reason of rejection or lack of rejection, since multiple factors not included in the null and alternative hypotheses can actually play a role. Furthermore, because of the different availability of flood observations all over the world, it is crucial to assess the impacts of varying record lengths on NFFA for practical reasons, and our research can provide technical guidance for engineers and policymakers.

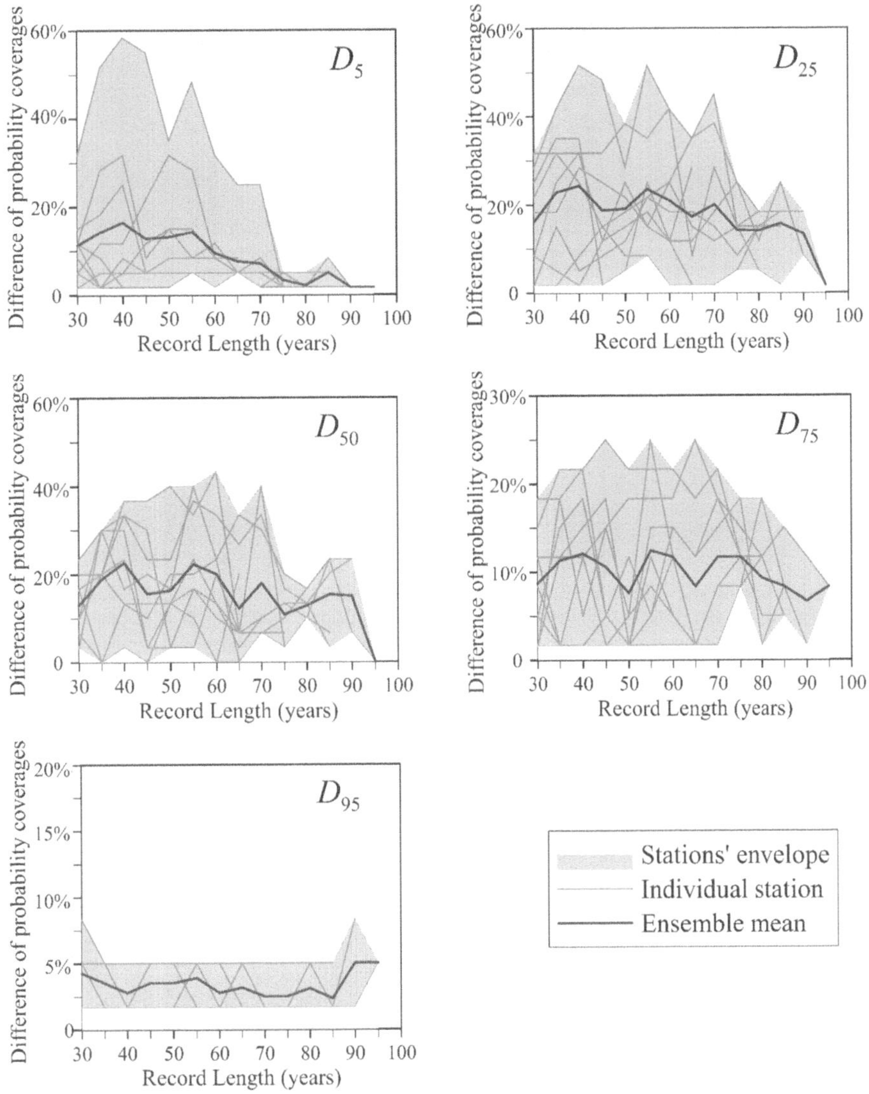

Figure 8 | The difference between theoretical probability coverage and actual probability coverages and the percentage of flood observations under 5th, 25th, 50th, 75th and 95th percentiles, denoted by D_5, D_{25}, D_{50}, D_{75} and D_{95}, respectively, using GEV-NS models. The difference was calculated using Equation (8), and ensemble mean was the average difference of nine stations.

The results of this study may be affected by the employed NS model. NS models make sense if we know *a priori* mechanism of change, and therefore we can set up parameters varying accordingly. Otherwise, we need to consider the model uncertainty including a variety of possible options and combinations in terms of covariates and link functions. Theoretically, all the three statistical parameters can be time-varying, but the shape parameter is sensitive and difficult to estimate. In practice, either location or scale parameters are likely to change under a changing climate, but in this study, only the location parameter of GEV-NS0 is time-varying to facilitate the model construction and investigation procedures. Although this kind of simplified GEV-NS0 model may reduce the model set and neglect an infinite set of possible alternative NS models, the aim of this study is not to select the optimal model for a selected basin but to investigate the impacts of record lengths on NFFA. It is a trade-off choice between the complexity of model structure of the NS model and the practical applications of NS model in this study. Besides, to strengthen the physical mechanism of NS models, researchers have recommended the use of covariates more relevant to flood process in the construction of NS models to the regions where human activity is an important contributor. In this study, we used time as a covariate for two reasons: (1) to focus the study on the main objectives, which are to examine the impacts of flood record lengths on the NS and ST design floods and associated uncertainties, and to explore the probable requirements of flood record length in NFFA and (2) human activities on the selected Norwegian catchments are not

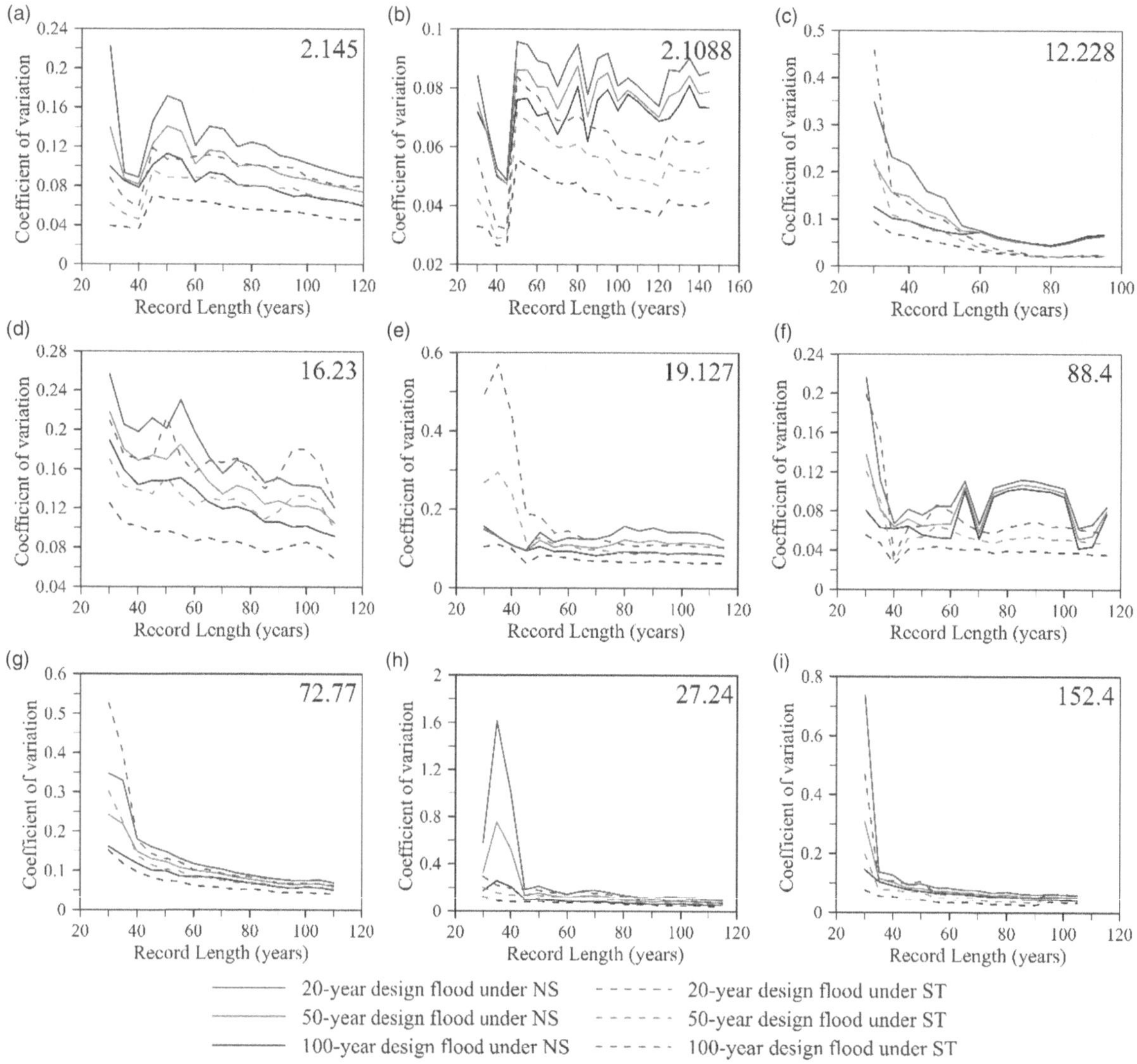

Figure 9 | Evolution of the CVs of 20-, 50- and 100-year design flood with the increase of record lengths under both ST (dashed lines) and NS (solid lines) conditions.

significant and the used maximum extreme values are normally less affected by non-climate factors compared with daily values or low-flow values. From the perspective of feasibility, in this study, all the record lengths of selected stations are larger than 100 years. So, it is very difficult to obtain physical covariates with such a long record length. It should be strengthened that time covariate is just a simple reflection of the gradual change of time series, and it lacks physical meaning. Thus, in future, when more observations data are available, it is feasible to incorporate physical covariates which are relevant to the flood process in the NS modeling.

Theoretically, the required sample size for a given analysis depends on the target statistic of interest (e.g., 50-, 100- and 1,000-year design flood) and required accuracy. In practical flood design problems, hydrological engineers work with the available flood observations they have and account for the corresponding uncertainties to a targeted return period. The recommended minimum sample size in flood design in the handbook of hydrology is to guarantee the reliability of design results for cases with no specific degrees of uncertainty in advance, and urge the hydrological engineers to collect more samples or turn to 'peaks over threshold' sampling scheme to reduce the huge uncertainty when estimating 1,000-year design flood using limited observations. Based on the results of this study, a minimum of 60 years of observations is recommended for NFFA to guarantee the superiority of the NS models. It should be mentioned that the threshold of 60-year observations switching from FFA to NFFA is not a 'magic' number. It is inferred based on fitting quality, predictive capabilities and the stability of GEV-

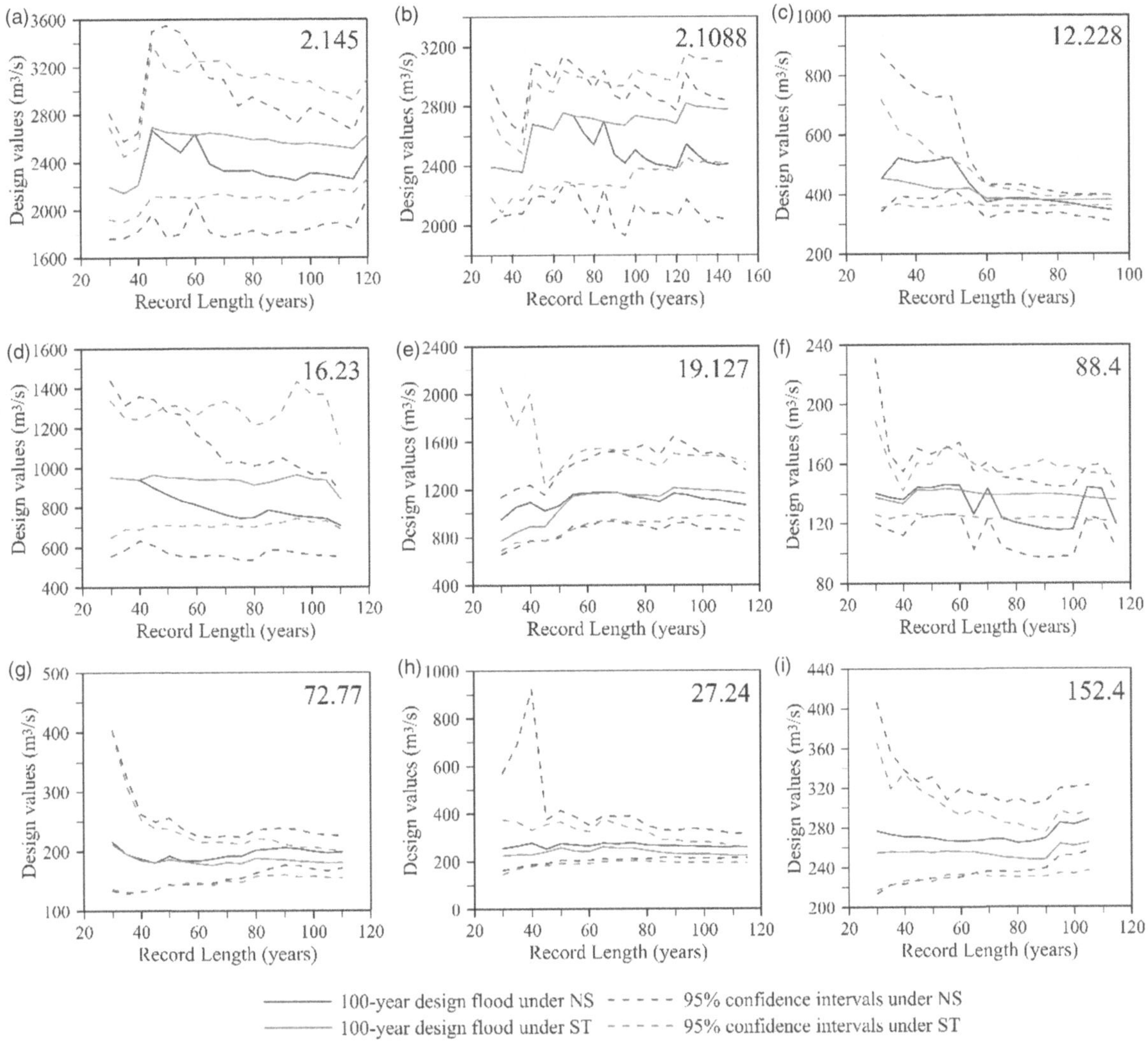

Figure 10 | Evolution of the 100-year design flood (solid lines) and associated CIs (dashed lines) with the increase of record lengths under both ST (red lines) and NS (blue lines) conditions. Please refer to the online version of this paper to see this figure in color: http://dx.doi.org/10.2166/nh.2021.026.

NS0 for most selected stations in Norway for estimating 20-, 50- and 100-year design flood, using GEV-NS0, which is an estimation that may depend on the nature of the time series, its memory, its distribution, the models used and the level of accuracy required. With that said, we still think that the recommendation of a plausible standard to guide the estimation of design flood for engineers, particularly for cases with limited observations, is beneficial for pushing forward the practical application of NS frequency analysis. Besides, it should be noted that in the practical design procedure, no matter whether the available observations satisfy the minimum requirement of record length, performing a cost–benefit (multicriteria) analysis considering the range of uncertainty corresponding to a predefined design value, thus getting the solution that optimizes the required criteria (i.e., structural and economic) is important. Since a number of other variables that are not specified in this study may affect the required sample size. For example, if we improve the level of required accuracy of design values in a specific problem, a series with 60 observations is likely to be not enough, or if we use the NS model with a different model structure whose location parameter and scale parameter are both time-varying, the required sample size may be different. Besides, in this study, only two stations' record lengths are larger than 120 years, influencing the analysis of probability coverage of GEV-ST and GEV-NS0. With the increase of observations in future, to promote the development of NFFA, further studies concerning the record length required for NFFA in other basins worldwide should be carried out.

5. CONCLUSIONS

The aim of this study was to evaluate the influence of record length on the ST and NS FFA. This aim was achieved using the split-sample testing for the stations with long record lengths (larger than 100 years), and several goodness-of-fit measures were used to evaluate the model performance and predictive capabilities of statistical models. Subsequently, the ADLL design method was employed to predict NS design floods, and the NNB was used to calculate CIs by resampling the original flood observations for record lengths varying from 30 to 145 years.

Based on the aforementioned results, the main conclusions of this study are as follows:

1. For frequency analysis of NS flood extremes, the performance of GEV-NS0 becomes better with the increase of record length and particularly they outperform GEV-ST models when record length is larger than 60 years in the selected basins in Norway, indicating the necessity of collecting more flood samples in NFFA rather than the minimum requirement of 30 samples in traditional FFA.
2. The predictive capabilities of extreme events (95th and 5th percentiles) are good for both GEV-ST and GEV-NS0 in the selected basins in Norway. However, compared with GEV-ST models, the difference between theoretical and actual probabilistic coverages becomes smaller, as record lengths increase for GEV-NS0. More importantly, the likelihood values of GEV-NS0 are larger than those of GEV-ST models when the record length is larger than 60 years for most stations, indicating the better predictive capabilities of GEV-NS0. The improved model performance and predictive capabilities of GEV-NS0 are thought to be attributed to their better capture of the variability of extreme flood events when more flood samples are added into the flood population.
3. The stability of the established GEV-ST and GEV-NS0 is improved, as record lengths increase since the CVs of 20-, 50- and 100-year design floods become smaller with the increase of record length. Particularly, the fluctuation of CV becomes gentle when the record length is roughly larger than 50–60 years for most stations in the selected basins. Besides, the CVs of GEV-ST are smaller than those of GEV-NS0.
4. The difference of 100-year design flood estimated using the ADLL method between GEV-ST and GEV-NS0 is relatively small or even becomes smaller when record length is larger than 50–60 years. In addition, the bootstrapped 95% CIs of the 100-year design flood become smaller with the increase of record length for both GEV-ST and GEV-NS0 models in the selected basins in Norway.

ACKNOWLEDGEMENTS

This study is financially supported jointly by the National Natural Science Foundation of China (Nos. 51525902 and 51909053), the Research Council of Norway (FRINATEK Project No. 274310 and KLIMAFORSK Project No. 302457), the Ministry of Education 'Plan 111' Fund of China (No. B18037), the Natural Science Foundation of Hebei Province (No. E2019402076), the Open Research Fund Program of State Key Laboratory of Water Resources and Hydropower Engineering Science (No. 2019SWG02) and the Youth Foundation of Education Department of Hebei Province (No. QN2019132), all of which are greatly appreciated.

CONFLICT OF INTEREST

The authors declare there is no conflict of interest regarding the publication of this paper.

DATA AVAILABILITY STATEMENT

Data cannot be made publicly available; readers should contact the corresponding author for details.

REFERENCES

Acero, F. J., Parey, S., Hoang, T. T. H., Dacunha-Castelle, D., García, J. A. & Gallego, M. C. 2017 Non-stationary future return levels for extreme rainfall over Extremadura (southwestern Iberian Peninsula). *Hydrological Sciences Journal* **62** (9), 1394–1411. doi:10.1080/02626667.2017.1328559.

Acero, F. J., Parey, S., García, J. A. & Dacunha-Castelle, D. 2018 Return level estimation of extreme rainfall over the Iberian Peninsula: comparison of methods. *Water* **10** (2), 179. doi:10.3390/w10020179.

Akaike, H. 1974 A new look at the statistical model identification. *IEEE Transactions on Automatic Control* **19** (6), 716–723. doi:10.1109/TAC.1974.1100705.

Arnell, N. W. & Gosling, S. N. 2016 The impacts of climate change on river flood risk at the global scale. *Climatic Change* **134** (3), 387–401. doi:10.1007/s10584-014-1084-5.

Bayazit, M. 2015 Nonstationarity of hydrological records and recent trends in trend analysis: a state-of-the-art review. *Environmental Processes* **2** (3), 527–542. doi:10.1007/s40710-015-0081-7.

Byun, K. & Hamlet, A. F. 2020 A risk-based analytical framework for quantifying non-stationary flood risks and establishing infrastructure design standards in a changing environment. *Journal of Hydrology* **584**, 124575. doi:10.1016/j.jhydrol.2020.124575.

Cheng, L. & Aghakouchak, A. 2014 Nonstationary precipitation intensity-duration-frequency curves for infrastructure design in a changing climate. *Scientific Reports* **4** (1), 378–381. doi:10.1038/srep07093.

Cooley, D. 2013 Return periods and return levels under climate change. In: *Extremes in a Changing Climate* (AghaKouchak, A., Easterling, D., Hsu, K., Schubert, S. & Sorooshian, S., eds). Springer, Dordrecht, Netherlands, pp. 97–114. doi:10.1007/978-94-007-4479-0_4.

Debele, S. E., Strupczewski, W. G. & Bogdanowicz, E. 2017 A comparison of three approaches to non-stationary flood frequency analysis. *Acta Geophysica* **65** (4), 863–883. doi:10.1007/s11600-017-0071-4.

Du, T., Xiong, L., Xu, C.-Y., Gippel, C. J., Guo, S. & Liu, P. 2015 Return period and risk analysis of nonstationary low-flow series under climate change. *Journal of Hydrology* **527**, 234–250. doi:10.1016/j.jhydrol.2015.04.041.

Ekeu-Wei, I. T. 2018 Evaluation of hydrological data collection challenges and flood estimation uncertainties in Nigeria. *Environment and Natural Resources Research* **8**, 44–54. doi:10.5539/enrr.v8n2p44.

Engeland, K., Wilson, D., Borsányi, P., Roald, L. & Holmqvist, E. 2017 Use of historical data in flood frequency analysis: a case study for four catchments in Norway. *Hydrology Research* **49** (2), 466–486. doi:10.2166/nh.2017.069.

Fleig, A. K., Andreassen, L. M., Barfod, E., Haga, J., Haugen, L. E., Hisdal, H., Melvold, K. & Saloranta, T. 2013 *Norwegian Hydrological Reference Dataset for Climate Change.* Norwegian Water Resources and Energy Directorate (NVE). Rapport 2.

Hall, J., Arheimer, B., Borga, M., Brazdil, R., Claps, P., Kiss, A., Kjeldsen, T. R., Kriaučiūnienė, J., Kundzewicz, Z. W. & Lang, M. 2013 Understanding flood regime changes in Europe: a state-of-the-art assessment. *Hydrology and Earth System Sciences* **18** (7), 2735–2772. doi:10.5194/hess-18-2735-2014.

Hallegatte, S., Green, C., Nicholls, R. J. & Corfee-Morlot, J. 2013 Future flood losses in major coastal cities. *Nature Climate Change* **3** (9), 802–806. doi:10.1038/nclimate1979.

Hao, Z. & Singh, V. P. 2016 Review of dependence modeling in hydrology and water resources. *Progress in Physical Geography* **40** (4), 549–578. doi:10.1177/0309133316632460.

Hou, J., Li, B., Tong, Y., Ma, L., Ball, J., Luo, H., Liang, Q. & Xia, J. 2020 Cause analysis for a new type of devastating flash flood. *Hydrology Research* **51** (1), 1–16. doi:10.2166/nh.2019.091.

Hu, Y., Liang, Z., Liu, Y., Zeng, X. & Wang, D. 2015 Uncertainty assessment of estimation of hydrological design values. *Stochastic Environmental Research and Risk Assessment* **29** (2), 501–511. doi:10.1007/s00477-014-0979-z.

Hu, Y., Liang, Z., Singh, V. P., Zhang, X., Wang, J., Li, B. & Wang, H. 2018 Concept of equivalent reliability for estimating the design flood under non-stationary conditions. *Water Resources Management* **32** (3), 997–1011. doi:10.1007/s11269-017-1851-y.

Hu, L., Nikolopoulos, E. I., Marra, F. & Anagnostou, E. N. 2020 Sensitivity of flood frequency analysis to data record, statistical model, and parameter estimation methods: an evaluation over the contiguous United States. *Journal of Flood Risk Management* **13** (1), e12580. doi:10.1111/jfr3.12580.

Jiang, C., Xiong, L., Yan, L., Dong, J. & Xu, C.-Y. 2019 Multivariate hydrologic design methods under nonstationary conditions and application to engineering practice. *Hydrology and Earth System Sciences* **23** (3), 1683–1704. doi:10.5194/hess-23-1683-2019.

Kan, G., Zhang, M., Liang, K., Wang, H., Jiang, Y., Li, J., Ding, L., He, X., Hong, Y., Zuo, D., Bao, Z. & Li, C. 2016 Improving water quantity simulation & forecasting to solve the energy-water-food nexus issue by using heterogeneous computing accelerated global optimization method. *Applied Energy* **210**, 420–433. doi:10.1016/j.apenergy.2016.08.017.

Khaliq, M. N., Ouarda, T. B. M. J., Ondo, J. C., Gachon, P. & Bobée, B. 2006 Frequency analysis of a sequence of dependent and/or non-stationary hydro-meteorological observations: a review. *Journal of Hydrology* **329** (3–4), 534–552. doi:10.1016/j.jhydrol.2006.03.004.

Kjeldsen, T. R., Macdonald, N., Lang, M., Mediero, L., Albuquerque, T., Bogdanowicz, E., Brázdil, R., Castellarin, A., David, V., Fleig, A., Gül, G. O., Kriauciuniene, J., Kohnová, S., Merz, B., Nicholson, O., Roald, L. A., Salinas, J. L., Sarauskiene, D., Šraj, M., Strupczewski, W., Szolgay, J., Toumazis, A., Vanneuville, W., Veijalainen, N. & Wilson, D. 2014 Documentary evidence of past floods in Europe and their utility in flood frequency estimation. *Journal of Hydrology* **517**, 963–973. doi:10.1016/j.jhydrol.2014.06.038.

Kleivane, I. 2006 *Vassføringsstasjonar på Vestlandet.* Norwegian Water Resources and Energy Directorate (NVE). Rapport 17.

Kobierska, F., Engeland, K. & Thorarinsdottir, T. L. 2017 Evaluation of design flood estimates – a case study for Norway. *Hydrology Research* **49** (2), 450–465. doi:10.2166/nh.2017.068.

Kochanek, K., Strupczewski, W. G., Bogdanowicz, E. & Markiewicz, I. 2020 The bias of the maximum likelihood estimates of flood quantiles based solely on the largest historical records. *Journal of Hydrology* **584**, 124740. doi:10.1016/j.jhydrol.2020.124740.

Kuang, Y., Xiong, L., Yu, K., Liu, P., Xu, C.-Y. & Yan, L. 2018 Comparison of first-order and second-order derived moment approaches in estimating annual runoff distribution. *Journal of Hydrologic Engineering* **23** (8), 04018034. doi:10.1061/(ASCE)HE.1943-5584.0001683.

Li, H., Gao, H., Zhou, Y., Xu, C.-Y., Ortega, M., & Sælthun, R. Z. & R, N. 2020 Usage of SIMWE model to model urban overland flood: a case study in Oslo. *Hydrology Research* **51** (2), 366–380. doi:10.2166/nh.2020.068.

Lovdata 2010 *Dam Safety Regulation (Damsikkerhetsforskriften).* Norway. Available from: https://lovdata.no/dokument/SF/forskrift/2009-12-18-1600.

Lu, M., Xu, Y., Shan, N., Wang, Q., Yuan, J. & Wang, J. 2019 Effect of urbanisation on extreme precipitation based on nonstationary models in the Yangtze River Delta metropolitan region. *Science of the Total Environment* **673**, 64–73. doi:10.1016/j.scitotenv.2019.03.413.

Lu, F., Song, X., Xiao, W., Zhu, K. & Xie, Z. 2020 Detecting the impact of climate and reservoirs on extreme floods using nonstationary frequency models. *Stochastic Environmental Research and Risk Assessment* **34** (1), 169–182. doi:10.1007/s00477-019-01747-2.

Luke, A., Vrugt, J. A., AghaKouchak, A., Matthew, R. & Sanders, B. F. 2017 Predicting nonstationary flood frequencies: evidence supports an updated stationarity thesis in the United States. *Water Resources Research* **53** (7), 5469–5494. doi:10.1002/2016WR019676.

Machado, M. J., Botero, B. A., López, J., Francés, F., Díez-Herrero, A. & Benito, G. 2015 Flood frequency analysis of historical flood data under stationary and non-stationary modelling. *Hydrology and Earth System Sciences* **19** (6), 2561–2576. doi:10.5194/hess-19-2561-2015.

Madsen, H., Lawrence, D., Lang, M., Martinkova, M. & Kjeldsen, T. R. 2014 Review of trend analysis and climate change projections of extreme precipitation and floods in Europe. *Journal of Hydrology* **519**, 3634–3650. doi:10.1016/j.jhydrol.2014.11.003.

McCuen, R. H. & Galloway, K. E. 2010 Record length requirements for annual maximum flood series. *Journal of Hydrologic Engineering* **15** (9), 704–707. doi:10.1061/(ASCE)HE.1943-5584.0000223.

Midttømme, G., Pettersson, L. E., Holmqvist, E., Nøtsund, Ø., Hisdal, H. & Sivertsgård, R. 2011 *Guidelines for Flood Estimation.* NVE retningslinjer, Oslo, Norway.

Milly, P. C. D., Betancourt, J., Falkenmark, M., Hirsch, R. M., Kundzewicz, Z. W., Lettenmaier, D. P. & Stouffer, R. J. 2008 Stationarity is dead: whither water management? *Science* **319** (5863), 573–574.

Milly, P. C. D., Betancourt, J., Falkenmark, M., Hirsch, R. M., Kundzewicz, Z. W., Lettenmaier, D. P., Stouffer, R. J., Dettinger, M. D. & Krysanova, V. 2015 On critiques of 'stationarity is dead: whither water management?'. *Water Resources Research* **51** (9), 7785–7789. doi:10.1002/2015WR017408.

Montanari, A., Young, G., Savenije, H. H. G., Hughes, D., Wagener, T., Ren, L. L., Koutsoyiannis, D., Cudennec, C., Toth, E. & Grimaldi, S. 2013 'Panta Rhei – Everything Flows': change in hydrology and society – the IAHS Scientific Decade 2013–2022. *Hydrological Sciences Journal* **58** (6), 1256–1275. doi:10.1080/02626667.2013.809088.

Obeysekera, J. & Salas, J. 2014 Quantifying the uncertainty of design floods under nonstationary conditions. *Journal of Hydrologic Engineering* **19** (7), 1438–1446. doi:10.1061/(ASCE)HE.1943-5584.0000931.

Olsen, R., Lambert, J. H. & Haimes, Y. Y. 1998 Risk of extreme events under nonstationary conditions. *Risk Analysis* **18**, 497–510. doi:10.1111/j.1539-6924.1998.tb00364.x.

Parey, S., Malek, F., Laurent, C. & Dacunha-Castelle, D. 2007 Trends and climate evolution: statistical approach for very high temperatures in France. *Climatic Change* **81** (3–4), 331–352. doi:10.1007/s10584-006-9116-4.

Parey, S., Hoang, T. T. H. & Dacunha-Castelle, D. 2010 Different ways to compute temperature return levels in the climate change context. *Environmetrics* **21** (7–8), 698–718. doi:10.1002/env.1060.

Pettersson, L. E. 2003 *Norges Hydrologiske Stasjonsnett.* Rapport 7. Norwegian Water Resources and Energy Directorate (NVE).

Qu, C., Li, J., Yan, L., Yan, P. & Lu, D. 2020 Non-stationary flood frequency analysis using cubic B-spline-based GAMLSS model. *Water* **12** (7), 1867. doi:10.3390/w12071867.

Read, L. K. & Vogel, R. M. 2015 Reliability, return periods, and risk under nonstationarity. *Water Resources Research* **51** (8), 6381–6398. doi:10.1002/2015WR017089.

Rigby, R. A. & Stasinopoulos, D. 2005 Generalized additive models for location, scale and shape. *Journal of the Royal Statistical Society: Series C (Applied Statistics)* **54** (3), 507–554. doi:10.1111/j.1467-9876.2005.00510.x.

Rootzén, H. & Katz, R. W. 2013 Design life level: quantifying risk in a changing climate. *Water Resources Research* **49** (9), 5964–5972. doi:10.1002/wrcr.20425.

Salas, J. D. & Obeysekera, J. 2014 Revisiting the concepts of return period and risk for nonstationary hydrologic extreme events. *Journal of Hydrologic Engineering* **19** (3), 554–568. doi:10.1061/(ASCE)HE.1943-5584.0000820.

Salas, J. D., Obeysekera, J. & Vogel, R. M. 2018 Techniques for assessing water infrastructure for nonstationary extreme events: a review. *Hydrological Sciences Journal* **63** (3), 325–352. doi:10.1080/02626667.2018.1426858.

Sarhadi, A. & Soulis, E. D. 2017 Time-varying extreme rainfall intensity-duration-frequency curves in a changing climate. *Geophysical Research Letters* **44** (5), 2454–2463. https://doi.org/10.1002/2016GL072201.

Sen, S., He, J. & Kasiviswanathan, K. S. 2020 Uncertainty quantification using the particle filter for non-stationary hydrological frequency analysis. *Journal of Hydrology* **584**, 124666. doi:10.1016/j.jhydrol.2020.124666.

Serago, J. M. & Vogel, R. M. 2018 Parsimonious nonstationary flood frequency analysis. *Advances in Water Resources* **112**, 1–16. doi:10.1016/j.advwatres.2017.11.026.

Serinaldi, F. & Kilsby, C. G. 2015 Stationarity is undead: uncertainty dominates the distribution of extremes. *Advances in Water Resources* **77**, 17–36. doi:10.1016/j.advwatres.2014.12.013.

Serinaldi, F. & Kilsby, C. G. 2016 The importance of prewhitening in change point analysis under persistence. *Stochastic Environmental Research and Risk Assessment* **30**, 763–777. doi:10.1007/s00477-015-1041-5.

Serinaldi, F., Kilsby, C. G. & Lombardo, F. 2018 Untenable nonstationarity: an assessment of the fitness for purpose of trend tests in hydrology. *Advances in Water Resources* **111**, 132–155. doi:10.1016/j.advwatres.2017.10.015.

Serinaldi, F., Chebana, F. & Kilsby, C. G. 2020 Dissecting innovative trend analysis. *Stochastic Environmental Research and Risk Assessment* **34** (5), 733–754. doi:10.1007/s00477-020-01797-x.

Slater, L. J. & Villarini, G. 2016 Recent trends in US flood risk. *Geophysical Research Letters* **43** (24), 12–428. doi:10.1002/2016GL071199.

Stedinger, J. R., Vogel, R. M. & Foufoula-Georgiou, E. 1993 Frequency analysis of extreme events. In: *Handbook of Hydrology* (Maidment, D. R., ed.). McGraw-Hill, New York.

Strupczewski, W. G., Kochanek, K. & Bogdanowicz, E. 2017 Historical floods in flood frequency analysis: is this game worth the candle? *Journal of Hydrology* **554**, 800–816. doi:10.1016/j.jhydrol.2017.09.034.

Su, C. & Chen, X. 2019 Assessing the effects of reservoirs on extreme flows using nonstationary flood frequency models with the modified reservoir index as a covariate. *Advances in Water Resources* **124**, 29–40. doi:10.1016/j.advwatres.2018.12.004.

Sun, P., Wen, Q., Zhang, Q., Singh, V. P., Sun, Y. & Li, J. 2018 Nonstationarity-based evaluation of flood frequency and flood risk in the Huai River basin, China. *Journal of Hydrology* **567**, 393–404. doi:10.1016/j.jhydrol.2018.10.031.

TEK10 2016 *Building Regulations (Byggtekninsk forskrift)*. Available from: https://dibk.no/byggeregler/tek/.

Um, M., Kim, Y., Markus, M. & Wuebbles, D. J. 2017 Modeling nonstationary extreme value distributions with nonlinear functions: an application using multiple precipitation projections for U.S. cities. *Journal of Hydrology* **552**, 396–406. doi:10.1016/j.jhydrol.2017.07.007.

Venvik, G., Bang-Kittilsen, A. & Boogaard, F. C. 2020 Risk assessment for areas prone to flooding and subsidence – a concept model with case study from Bergen, Western Norway. *Hydrology Research* **51** (2), 322–338. doi:10.2166/nh.2019.030.

Villarini, G., Smith, J. A., Serinaldi, F., Bales, J., Bates, P. D. & Krajewski, W. F. 2009 Flood frequency analysis for nonstationary annual peak records in an urban drainage basin. *Advances in Water Resources* **32** (8), 1255–1266. doi:10.1016/j.advwatres.2009.05.003.

Vogel, R. M., Yaindl, C. & Walter, M. 2011 Nonstationarity: flood magnification and recurrence reduction factors in the United States1. *Journal of the American Water Resources Association* **47** (3), 464–474. doi:10.1111/j.1752-1688.2011.00541.x.

Vormoor, K., Lawrence, D., Heistermann, M. & Bronstert, A. 2015 Climate change impacts on the seasonality and generation processes of floods – projections and uncertainties for catchments with mixed snowmelt/rainfall regimes. *Hydrology and Earth System Sciences* **19** (2), 913–931. doi:10.5194/hess-19-913-2015.

Vormoor, K., Lawrence, D., Schlichting, L., Wilson, D. & Wong, W. K. 2016 Evidence for changes in the magnitude and frequency of observed rainfall vs. snowmelt driven floods in Norway. *Journal of Hydrology* **538**, 33–48. doi:10.1016/j.jhydrol.2016.03.066.

Wang, J., Xu, Y., Wang, Y., Yuan, J., Wang, Q. & Xiang, J. 2019 Non-stationarity analysis of extreme water level in response to climate change and urbanization in the Taihu Basin, China. *Stochastic Environmental Research and Risk Assessment* **33** (3), 891–904. doi:10.1007/s00477-019-01666-2.

Westra, S., Alexander, L. V. & Zwiers, F. W. 2013 Global increasing trends in annual maximum daily precipitation. *Journal of Climate* **26** (11), 3904–3918. doi:10.1175/JCLI-D-12-00502.1.

Xie, P., Wu, Z., Sang, Y., Gu, H., Zhao, Y. & Singh, V. P. 2018 Evaluation of the significance of abrupt changes in precipitation and runoff process in China. *Journal of Hydrology* **560**, 451–460. doi:10.1016/j.jhydrol.2018.02.036.

Xiong, L., Yan, L., Du, T., Yan, P., Li, L. & Xu, W. 2019 Impacts of climate change on urban extreme rainfall and drainage infrastructure performance: a case study in Wuhan City, China. *Irrigation and Drainage* **68** (2), 152–164. doi:10.1002/ird.2316.

Xiong, B., Xiong, L., Guo, S., Xu, C.-Y., Xia, J., Zhong, Y. & Yang, H. 2020 Nonstationary frequency analysis of censored data: a case study of the floods in the Yangtze River from 1470 to 2017. *Water Resources Research* **56**, e2020WR027112. doi:10.1029/2020WR027112.

Xu, W., Jiang, C., Yan, L., Li, L. & Liu, S. 2018 An adaptive metropolis-hastings optimization algorithm of Bayesian estimation in non-stationary flood frequency analysis. *Water Resources Management* **32** (4), 1343–1366. doi:10.1007/s11269-017-1873-5.

Yan, L., Xiong, L., Guo, S., Xu, C.-Y., Xia, J. & Du, T. 2017a Comparison of four nonstationary hydrologic design methods for changing environment. *Journal of Hydrology* **551**, 132–150. doi:10.1016/j.jhydrol.2017.06.001.

Yan, L., Xiong, L., Liu, D., Hu, T. & Xu, C.-Y. 2017b Frequency analysis of nonstationary annual maximum flood series using the time-varying two-component mixture distributions. *Hydrological Processes* **31** (1), 69–89. doi:10.1002/hyp.10965.

Yan, L., Li, L., Yan, P., He, H., Li, J. & Lu, D. 2019a Nonstationary flood hazard analysis in response to climate change and population growth. *Water* **11** (9), 1811. doi:10.3390/w11091811.

Yan, L., Xiong, L., Ruan, G., Xu, C.-Y., Yan, P. & Liu, P. 2019b Reducing uncertainty of design floods of two-component mixture distributions by utilizing flood timescale to classify flood types in seasonally snow covered region. *Journal of Hydrology* **574**, 588–608. doi:10.1016/j.jhydrol.2019.04.056.

Yan, L., Xiong, L., Luan, Q., Jiang, C., Yu, K. & Xu, C.-Y. 2020 On the applicability of the expected waiting time method in nonstationary flood design. *Water Resources Management* **34** (8), 2585–2601. doi:10.1007/s11269-020-02581-w.

Yan, L., Xiong, L., Jiang, C., Zhang, M., Wang, D. & Xu, C.-Y. 2021 Updating intensity–duration–frequency curves for urban infrastructure design under a changing environment. *WIREs Water* e1519. doi:10.1002/wat2.1519.

Yin, J., Gentine, P., Zhou, S., Sullivan, S. C., Wang, R., Zhang, Y. & Guo, S. 2018 Large increase in global storm runoff extremes driven by climate and anthropogenic changes. *Nature Communications* **9** (1), 4389. doi:10.1038/s41467-018-06765-2.

Yue, S., Pilon, P., Phinney, B. & Cavadias, G. 2002 The influence of autocorrelation on the ability to detect trend in hydrological series. *Hydrological Processes* **16** (9), 1807–1829. doi:10.1002/hyp.1095.

Zhang, W., Villarini, G., Vecchi, G. A. & Smith, J. A. 2018 Urbanization exacerbated the rainfall and flooding caused by hurricane Harvey in Houston. *Nature* **563** (7731), 384–388. doi:10.1038/s41586-018-0676-z.

Zhang, X., Dong, Q., Costa, V. & Wang, X. 2019 A hierarchical Bayesian model for decomposing the impacts of human activities and climate change on water resources in China. *Science of the Total Environment* **665**, 836–847. doi:10.1016/j.scitotenv.2019.02.189.

Zwiers, F. W., Zhang, X. & Feng, Y. 2010 Anthropogenic influence on long return period daily temperature extremes at regional scales. *Journal of Climate* **24** (3), 881–892. doi:10.1175/2010JCLI3908.1.

First received 19 March 2021; accepted in revised form 5 July 2021. Available online 14 July 2021

doi: 10.2166/nh.2021.061

Decline in net primary productivity caused by severe droughts: evidence from the Pearl River basin in China

Yuliang Zhou* and Ping Zhou
College of Civil Engineering, Hefei University of Technology, Hefei 230009, China
*Corresponding author. E-mail: ZYL54600@163.com

ABSTRACT

Understanding the spatiotemporal characteristics of drought events and their impacts on terrestrial net primary productivity (NPP) is crucial for drought mitigation and environmental protection. This study, by taking the Pearl River basin as the case region, investigated drought duration, severity, intensity, affected area, and centroids during 1960–2015 based on the Standardized Evapotranspiration Deficit Index and three-dimensional clustering algorithm and then revealed how these drought characteristics have affected NPP. Results showed that there were altogether 32 severe drought events lasting at least 3 months in the basin, with half lasting longer than 6 months. The total NPP loss significantly correlated with drought severity and intensity. Most drought events caused a reduction in NPP across more than half of the drought-affected area; specifically, the February–December drought in 2011 has cut NPP by 31.85 Tg C, accounting for 11.7% of the regional annual mean NPP, while the September 2009–September 2010 drought caused a decrease of 20.26 Tg C in NPP. Our research improves the insight into the relationship between NPP and drought, which helps decision-makers manage droughts and provides guidance for drought-related studies across other regions.

Key words: drought events, net primary productivity, Pearl River basin, SEDI, spatiotemporal characteristics

HIGHLIGHTS

- Most drought events identified in the Pearl River basin from 1960 to 2015 lasted for more than 3 months.
- The total NPP loss significantly correlated with drought severity and intensity.
- The February–December 2011 drought in the PRB reduced NPP by 31.85 Tg C.

GRAPHICAL ABSTRACT

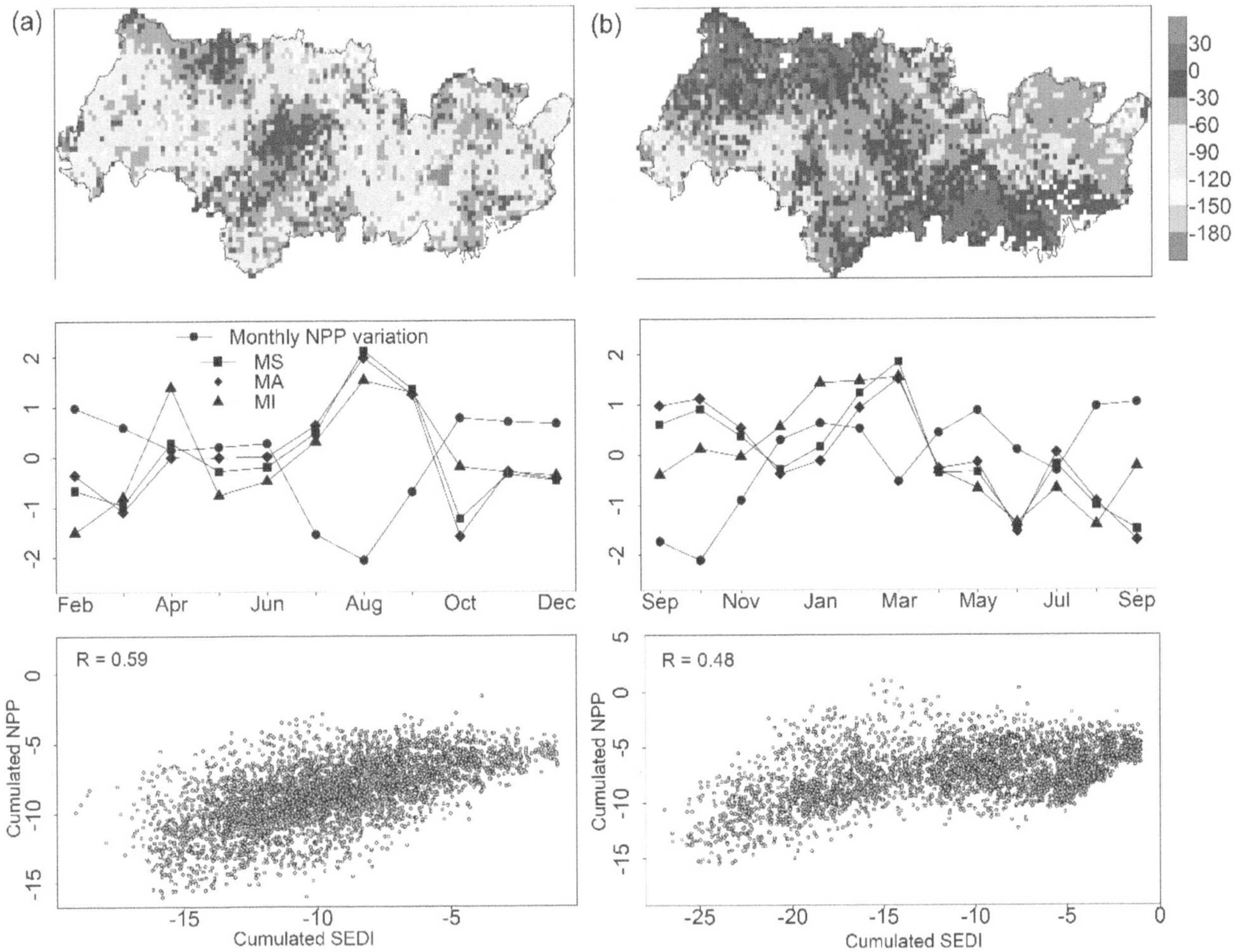

INTRODUCTION

Drought disaster is considered as one of the most complex natural disasters with devastating impacts on agriculture, economy, water resources, and ecosystems (Zhao & Running 2010; Lesk *et al.* 2016; Li *et al.* 2021a, 2021b). During the past decades, frequent and long-lasting droughts have occurred across many parts of the world, even in humid regions, leading to severe ecological environmental problems (Wu *et al.* 2017a; Tijdeman *et al.* 2020). Droughts in China are expected to increase particularly in south China due to global warming (Su *et al.* 2018; Li *et al.* 2021a, 2021b). Understanding the spatiotemporal characteristics of drought events and their impacts is of crucial importance for providing fundamental information on drought which in turn help prepare for the consequences.

To identify drought events and analyze their characteristics, several methods, including the run theory (Ayantobo *et al.* 2017), wavelet analysis, principle component analysis and cluster analysis (Gocic & Trajkovic 2014), have been commonly used. Among the methods, the run theory and the threshold level are robust ways of analyzing droughts (Li *et al.* 2020a), while the wavelet analysis is widely utilized to analyze drought periodicity and teleconnections (Belayneh *et al.* 2014). These methods provide useful references for understanding droughts. However, they do not fully extract space–time drought structures; what they consider is the temporal evolution within a fixed area or the spatial pattern over fixed durations (Lloyd-Hughes 2012; Xu *et al.* 2015). Consequently, much of the spatiotemporal drought information is reduced to a lower subspace (one or two dimensions), and the real drought structures in space–time dimensions are discarded (Sheffield *et al.* 2009; Guo *et al.* 2018). In fact, drought is a three-dimensional phenomenon that simultaneously evolves in both time and space (Li *et al.*

2020a). Until more recently, the three-dimensional clustering algorithm is applied to drought analysis at global and regional scales (Herrera-Estrada *et al.* 2017; Liu *et al.* 2019).

There is still little attention paid to the impacts of droughts on terrestrial ecosystems. As one of the major climatological disasters, drought can have significant impacts on terrestrial ecosystems, especially for the terrestrial net primary productivity (NPP) (Anderegg *et al.* 2015). As the initial step in the terrestrial carbon cycle, NPP represents the amount of atmospheric carbon stored by plants and accumulated as biomass, which plays a key role in the global carbon cycle and climate change (Piao *et al.* 2009). NPP is also considered as a key variable for the evaluation of effects of drought events on ecosystem conditions due to its sensitivity to droughts (Zhao & Running 2010). Therefore, exploring the response of NPP to drought events is essential to understanding the impact of climate change on the regional carbon cycle (Liang *et al.* 2015). Chen *et al.* (2013) analyzed the impact of drought on NPP globally and found a strong positive relationship between global average moisture availability and NPP. Castro *et al.* (2018) investigated the effect of drought on productivity in a Costa Rican tropical dry forest, pointing out that primary productivity was reduced by 13 and 42% during the drought seasons in 2014 and 2015, respectively. Several studies have discussed the impact of the increasing droughts on national and regional NPP variations across China, providing key information on the potential biosphere feedback to drought risks (Sun *et al.* 2016). For example, Zhang *et al.* (2012) focused on the effect of the 2010 spring drought on NPP in Southwest China and reported that the drought has reduced NPP by 46 Tg C.

The Pearl River basin, as the second-largest drainage basin in China, is economically developed and is of great importance to the socioeconomic development of China (Wu *et al.* 2017b). With the recent economic boom, the basin has become the major carbon dioxide emitter in China (Zhang *et al.* 2015). As a typical humid region, the basin has an abundant number of vegetative ecosystems, providing great potential for local biological carbon sequestration to partially offset fossil fuel emissions. However, this region has suffered from frequent and severe droughts in recent decades (Deng *et al.* 2018); these disturbances may have large impacts on vegetative productivity and thus weaken the carbon sequestration of vegetation. Although the variations in droughts over this region have received increasing attention (Xiao *et al.* 2016; Deng *et al.* 2018), few studies have specifically focused on the impact of long-lasting and severe droughts on NPP.

To reveal how drought events have affected NPP, this study took the Pearl River basin as the case region, attempting to map the spatiotemporal characteristics of drought events from 1960 to 2015, and subsequently exploring the relationships between drought characteristics (e.g. severity and intensity) and NPP. Drought events were identified using the Standardized Evapotranspiration Deficit Index (SEDI) and the three-dimensional clustering algorithm (Lloyd-Hughes 2012; Xu *et al.* 2015), and the characteristics of drought events, including duration, severity, intensity, centroid, and affected areas, were investigated afterward. Finally, the impacts of droughts on NPP were assessed. We provide detailed information on drought evolution and an insight into the variations in terrestrial ecosystems caused by drought disturbance as well as the biosphere feedback to drought risks.

STUDY AREA AND DATA

Study area

The Pearl River basin (102°14′–115°53′E, 21°31′–26°49′N) (Figure 1) is the second-largest basin in China. It covers a total drainage area of 453,690 km^2 with three main tributaries including the West River, the East River, and the North River. The basin is dominated by a tropical and sub-tropical climate. Rainfall in this basin is unevenly distributed in time with precipitation during March–October accounting for about 90% of the annual total (Wu *et al.* 2018). In recent decades, droughts in the basin exhibit a spreading trend (Li *et al.* 2020c). For example, the upstream has experienced severe drought during September 2009 and May 2010. Moreover, due to rapid urban expansion and the population explosion, the basin suffers from increasing water use stress (Deng *et al.* 2018). A comprehensive understanding of the drought conditions is of great importance for dealing with the potential and enormous threats caused by droughts.

Data

Daily meteorological datasets from 1960 to 2015 were collected from 91 observational stations within and surrounding the basin (Figure 1). The datasets include precipitation, maximum air temperature, mean air temperature, minimum air temperature, air pressure, wind speed, relative humidity, and sunshine duration. They were provided by the China Meteorological Administration (http://www.cma.gov.cn/en2014/) and the Resources and Environmental Science Data Center, Chinese Academy of Sciences (http://www.resdc.cn/Default.aspx). The temporal coverage exceeded 99.7%, and the missing data

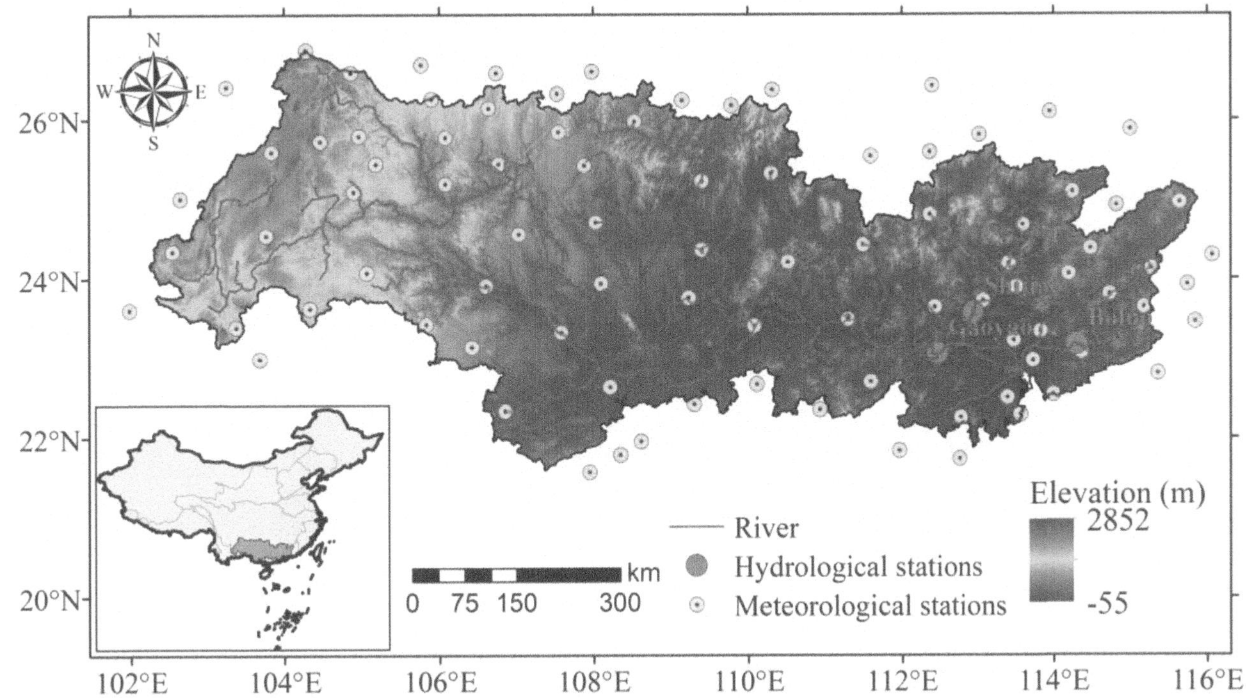

Figure 1 | The study area and the meteorological and hydrological stations.

were filled in with the average value of the neighboring days. The datasets were interpolated into a 0.1° spatial resolution using the kriging method.

Monthly streamflow observation data at the Gaoyao, Shijiao, and Boluo hydrological stations during 1960–2000 were collected to calibrate and validate the VIC model for the West, North, and East River sub-basins, respectively. All streamflow data were provided by the Hydrological Bureau of Guangdong Province. The elevation data with a spatial resolution of 90 × 90 m were downloaded from the International Scientific and Technical Data Mirror Site. The FAO's Harmonized World Soil Database with a 1 km spatial resolution was used for soil parameter specification in the VIC model. The hydrological attributes required by the VIC model were obtained using formulas developed by Saxton & Rawls (2006). To match the spatial scale, the soil data and elevation data were resampled to a 0.1° spatial resolution.

NPP datasets with a spatial resolution of 8 × 8 km from 1982 to 2015 were calculated from the Carnegie Ames Stanford Application Model by using the monthly normalized difference vegetation index (NDVI) and the meteorological data. The study of Lai *et al.* (2018), which used the NPP datasets, demonstrates that NPP data simulated by this model are satisfactory. The original NPP datasets were resampled to a 0.1° spatial resolution.

METHODS

VIC model

The VIC model was employed to compute evapotranspiration (ETa). It has been widely used to simulate soil moisture and evapotranspiration, and many studies have demonstrated that the VIC-based output such as soil moisture, streamflow, and evapotranspiration are reliable (Xia *et al.* 2014; Zhou *et al.* 2016). The model has been successfully applied in the Pearl River basin for soil moisture and streamflow simulations (Niu 2013; Wu *et al.* 2014). The simulated soil moisture and evapotranspiration are capable of capturing the variation of dry and wet conditions over the basin (Niu & Chen 2016). In this study, the VIC model was verified through a comparison between the simulated stream flows and the observations. The Nash–Sutcliffe efficiency (NSE), relative error (Bias), and determination coefficient (R^2) were used for the evaluation of model accuracy (Zhou *et al.* 2019, 2020; Kao *et al.* 2020; Lin *et al.* 2020). A significant coincidence rate with $P < 0.05$ was detected in both the calibration and validation (Figure 2), demonstrating that the VIC-based ETa is adaptable and rational for the subsequent drought analysis. More details on the calibration and validation can be found in Li *et al.* (2020b).

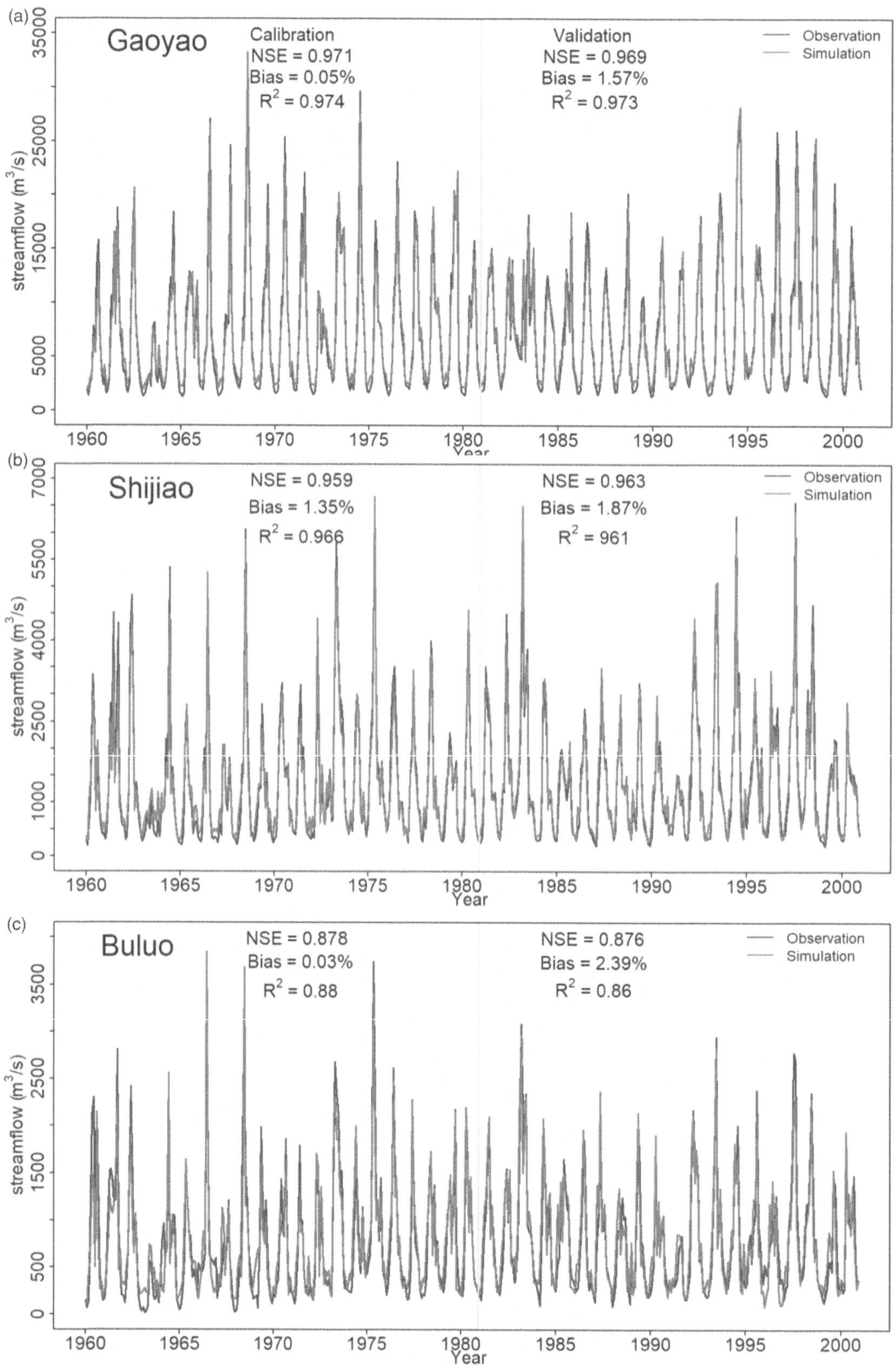

Figure 2 | Calibration and validation of the VIC model through comparison between the observed and simulated stream flows at the Gaoyao (a), Shijiao (b), and Boluo (c) hydrological stations during 1960–2000.

Drought index

Currently, the Standardized Precipitation Index (SPI) and the Standardized Precipitation Evapotranspiration Index (SPEI) are commonly applied to exploring the response of ecosystems to droughts. However, compared with precipitation-dependent indices, evapotranspiration-based drought indices may be more appropriate for the identification of ecosystem droughts (Vicente-Serrano *et al.* 2018). Therefore, the SEDI was developed (Vicente-Serrano *et al.* 2018) which is determined by the standardization of the evaporation deficit (ED) defined as the difference between ETa and atmospheric evaporative demand (AED). Vegetation activity is greatly influenced by ED, whereby high ED leads to stomatal closure reducing photosynthetic activity, carbohydrate accumulation, and NPP (Vicente-Serrano *et al.* 2015). If ED keeps increasing, the wilting point is reached leading to vegetation mortality due to excessive depletion of plant available water and carbohydrates (Anderegg *et al.* 2015). The SEDI considers plant growth and responding mechanisms and therefore is more suitable for studying the effect of drought on vegetation productivity (Vicente-Serrano *et al.* 2018). To validate the SEDI, it is appropriate to compare it with other drought indices. In this study, we not only compared it with SPEI, but also compared it with another widely used drought index, i.e. self-calibrating Palmer Drought Severity Index (scPDSI).

As stated above, we used the ETa calculated from the VIC model because *in situ* ETa measurement is usually unavailable. There are different ways to determine AED, including using pan evaporation or potential evapotranspiration (PET). In this study, AED was represented by PET. The Thornthwaite and Penman–Monteith equations are the two popular methods to calculate PET. However, the Thornthwaite equation tends to underestimate evapotranspiration, as it only considers temperature without considering other climate factors. Therefore, the Penman–Monteith equation was adopted in our study.

Drought event identification

In this study, SEDI was used to identify drought events. The identification is based on the three-dimensional clustering method (Lloyd-Hughes 2012). The three-dimensional space, including the longitude, latitude, and time, is built by stacking spatial SEDI maps over the basin from 1960 to 2015. Detailed steps are as follows.

First, drought patches were identified for each month. Drought condition was identified when its SEDI value was less than -1 at a grid cell. The density-based spatial clustering of application with noise clustering algorithm (Sander *et al.* 1998) was used to extract the drought patches, and the small and discontinuous patches were filtered out.

Secondly, a drought patch with an area greater than a prescribed threshold was selected. Drought clusters having small areas can persist for many years through tenuous spatial connectivity, and thus prolong drought duration (Sheffield *et al.* 2009). Previous studies suggest that a threshold of 1.6% of the total area of region of interest is suitable for drought identification (Xu *et al.* 2015; Guo *et al.* 2018). Therefore, we used the threshold of 7,259 km^2 (1.6% of the total area of Pearl River basin).

Thirdly, the overlap area between patches was calculated for two consecutive months. Two drought patches are considered to belong to a same event if the overlapping area is larger than the specified threshold (i.e. 7,259 km^2); otherwise, they are considered two independent events.

Finally, drought events were characterized by five parameters, including drought duration (DD), drought severity (DS), drought intensity (DI), drought area (DA), and drought centroid (DC). To analyze drought evolution, monthly severity (MS), intensity (MI), affected area (MA), and centroid (MC) during a drought event were also calculated. More details about this method can be found in Xu *et al.* (2015).

Detection of NPP responses to drought events

To assess NPP responses to large-scale and severe drought events, we focused on the NPP variation across the drought-affected area. The three-dimensional cluster of NPP was extracted with the same spatiotemporal dimensions of the drought events that occurred during 1982–2015. We analyzed NPP anomalies at each grid during a drought event. NPP anomaly caused by a drought event at each grid is calculated by the following equation:

$$\Delta \mathrm{NPP}(i, j, k) = \mathrm{NPP}(i, j, k) - \overline{\mathrm{NPP}}(i, j, k) \tag{1}$$

where $\Delta\mathrm{NPP}(i, j, k)$ refers to the monthly NPP anomaly within a grid during a drought event (g cm^{-2}); i and j indicate the grid location; k is the time (month); $\mathrm{NPP}(i, j, k)$ represents NPP for the kth month at the grid(i, j); $\overline{\mathrm{NPP}}$ is the mean NPP at the grid(i, j) for the kth month during the reference period of 1982–2015; positive/negative $\Delta\mathrm{NPP}(i, j, k)$ indicates change in

NPP at grid(i, j) for the kth month. Afterward, we calculated the total NPP anomaly (Tg C) (defined as the cumulative NPP anomaly for all grids during a drought event), and the positive/negative total NPP anomaly indicates increase/decrease in NPP. The total NPP anomaly during the nth drought event is defined as follows:

$$TNA_n = \sum_i \sum_j \sum_k (\Delta NPP(i, j, k) \times \text{area}(i, j, k)) \quad (2)$$

RESULTS

Validation of SEDI

A comparison between the SEDI and the multi-time scale SPEI and monthly scPDSI was carried out and shown in Figure 3. The SEDI series generally exhibited significant correlations ($P < 0.05$) with the short time scale SPEI; the correlations were greater than 0.4 for the 1- and 3-month SPEI across almost the entire region; however, the statistical significance of the correlations gradually decrease for longer time scales. A significant correlation between SEDI and scPDSI was also found for most regions. To further validate SEDI, soil moisture was employed to verify the identified drought events, given that soil moisture generally decreases during severe droughts (Xu *et al.* 2015; Guo *et al.* 2018). The three layers of soil moisture were used to verify droughts by fully considering the capability of SEDI to reflect soil moisture. Soil moisture standardized anomaly (SSMA) was first obtained for each grid using the z-score standardized method (Wu *et al.* 2001) to reduce the effects of seasonality and to facilitate the comparison (Guo *et al.* 2018). Using the three-dimensional clustering algorithm and the monthly SEDI time series from 1960 to 2015, the drought events were discerned and drought severities were computed. The corresponding drought events were also achieved based on the SSMA time series. Each SSMA cluster had the same

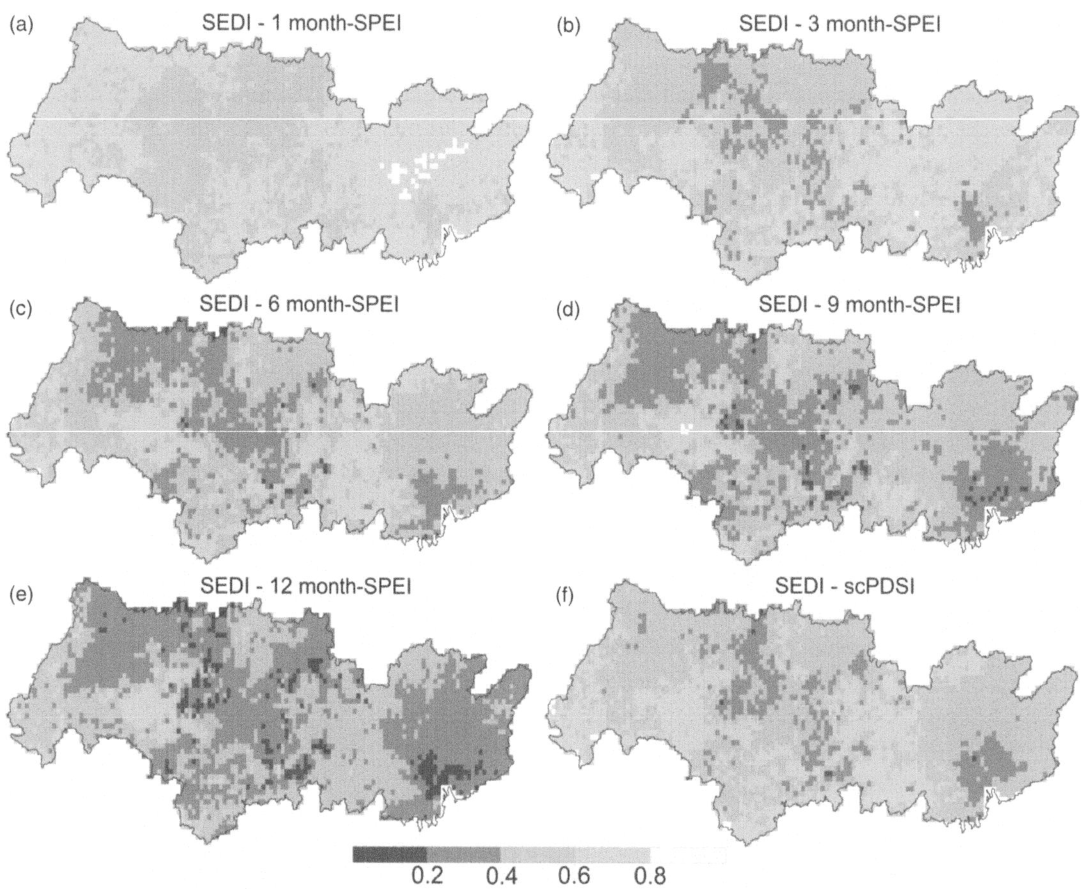

Figure 3 | Spatial distribution of coefficient correlation between SEDI and multi-time scale SPEI and monthly scPDSI.

spatiotemporal dimension with the drought event identified by SEDI to ensure that they had the same duration and affected area. The corresponding drought severities were calculated based on the SSMA clusters.

Figure 4 shows the comparison between drought severities calculated from SSMA and SEDI. The SEDI and the SSMA drought severities were highly correlated ($P < 0.01$); R^2 values were greater than 0.93, indicating that soil moisture in all three layers is sensitive to large-scale and severe drought events identified by SEDI. Despite the abundant soil moisture in humid areas, evapotranspiration was strong with lack of precipitation during drought periods, which has led to a decrease in soil moisture. In addition, humid regions such as southern China generally cover dense vegetation which would pump more water from the deep soil during severe droughts, resulting in a decrease in soil moisture (Wang *et al.* 2016). Overall, these analyses indicate that the SEDI can be used to identify drought events.

Spatiotemporal variation in drought events

A total of 32 prolonged and severe drought events lasting for at least 3 months were identified. All drought events (ranked by their severity) and their characteristics are shown in Table 1. Most of the drought events lasted for 3–6 months. Droughts lasting longer than 9 months occurred several times and only one lasted longer than 1 year. In addition, prolonged droughts mainly occurred during the past two decades. More specifically, the 1962–1963 drought event was the most severe of the past 56 years; it persisted for 11 months from December 1962 to October 1963, with a severity of 4.4×10^6 km^2/month, while the most prolonged drought lasted for 13 months from September 2009 to September 2010, ranking second in the list. Two drought events affected almost the whole basin, the drought area of which was larger than 400,000 km^2. Moreover, drought events with the same duration could have quite different drought-affected areas and severities. For example, although there were three drought events lasting for 11 months, the severity of the 2006–2007 event (the 9th event) was completely different from that which occurred during 1962–1963 (the 1st event) and that in 2011 (the 3rd event) due to different drought intensities and affected areas; its severity was relatively low, i.e. 1.72×10^6 km^2/month. Also, despite the similar severities, the duration and intensity can differ, as is seen in the 12th and 13th events; the 12th event had a relatively small region with low intensity but had a long duration, whereas the 13th event affected a larger area with a shorter duration and a higher intensity. On the other hand, the 6th drought event lasted for only 6 months but the severity was relatively high due to the high intensity and large affected area. On the whole, longer duration drought events had higher severities and larger affected areas, but droughts having short durations may have high severity.

Figure 5 shows the spatial distribution of the centroids for all drought events. The centroids of the prolonged and severe drought events were mainly detected in the upper and middle reaches. The drought events in the eastern part of the basin were generally of shorter duration and were less severe. The temporal distribution of the characteristics of the drought events is shown in Figure 6. It can be seen that drought events did not overlap before 2000, indicating that the drought events occurred separately over time. The basin experienced fewer severe droughts during the 1970s and 1980s compared

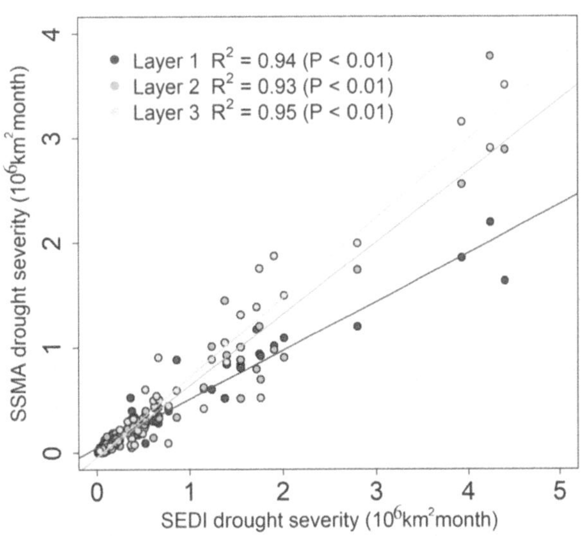

Figure 4 | Correlation between drought severities based on the SEDI and the normalized soil moisture for three layers.

Table 1 | The drought events and their characteristics arcoss the Pearl River basin during 1960–2015

Rank	Persistent time	DD (month)	DS (10^6 km^2·month)	DI	DA (10^5 km^2)	DC
1	1962.12–1963.10	11	4.40	0.97	4.10	109.42°E, 24.20°N
2	2009.09–2010.09	13	4.24	0.81	4.05	107.03°E, 24.49°N
3	2011.02–2011.12	11	3.93	0.87	4.09	109.03°E, 24.30°N
4	1998.09–1999.06	10	2.80	0.71	3.94	109.76°E, 24.12°N
5	2003.11–2004.08	10	2.01	0.54	3.72	110.00°E, 24.12°N
6	2005.09–2006.02	6	1.90	0.81	3.93	109.81°E, 24.16°N
7	2015.02–2015.11	10	1.76	0.48	3.63	107.44°E, 24.50°N
8	1960.01–1960.09	9	1.75	0.48	4.05	107.67°E, 24.53°N
9	2006.08–2007.06	11	1.72	0.46	3.41	107.34°E, 24.13°N
10	2007.08–2008.01	6	1.55	0.71	3.62	110.13°E, 24.07°N
11	1989.07–1989.12	6	1.54	0.70	3.68	109.40°E, 24.04°N
12	2012.08–2013.04	9	1.40	0.51	3.07	106.72°E, 24.53°N
13	1992.09–1993.01	5	1.38	0.75	3.69	108.73°E, 24.43°N
14	2004.10–2005.01	4	1.24	0.85	3.67	110.41°E, 23.83°N
15	1986.12–1987.06	7	1.16	0.46	3.57	109.37°E, 24.37°N
16	2013.07–2014.03	9	0.86	0.37	2.57	106.87°E, 24.92°N
17	2009.01–2009.03	3	0.77	0.84	3.05	110.83°E, 24.05°N
18	2005.03–2005.05	3	0.67	0.73	3.04	107.85°E, 24.10°N
19	2006.04–2006.06	3	0.66	0.83	2.65	107.09°E, 24.27°N
20	2012.02–2012.05	4	0.64	0.84	1.91	105.32°E, 24.58°N
21	1991.09–1991.11	3	0.62	1.00	2.05	110.92°E, 23.73°N
22	2000.07–2000.12	6	0.61	0.55	1.84	109.97°E, 23.27°N
23	2002.03–2002.07	5	0.52	0.87	1.19	113.50°E, 23.72°N
24	1964.12–1965.03	4	0.51	0.64	1.99	110.93°E, 24.11°N
25	1966.09–1966.11	3	0.49	0.68	2.39	111.48°E, 23.97°N
26	1977.03–1977.05	3	0.44	0.96	1.53	112.89°E, 23.73°N
27	1991.05–1991.07	3	0.39	0.74	1.76	112.18°E, 24.13°N
28	1980.11–1981.01	3	0.37	0.81	1.54	112.34°E, 23.66°N
29	1969.01–1969.06	6	0.37	0.42	1.49	105.73°E, 24.83°N
30	1980.03–1980.05	3	0.20	0.67	1.00	105.63°E, 23.70°N
31	1967.11–1968.01	3	0.13	0.69	0.65	114.27°E, 23.84°N
32	1979.02–1979.04	3	0.11	0.90	0.41	103.56°E, 24.86°N

to other periods, and the droughts during this period were less intense with shorter durations and smaller affected areas. After 2000, the basin experienced frequent large-scale droughts with higher severity and larger drought-affected areas. The time intervals between droughts were short, and almost all drought events had multiple peaks for both severity and affected area. Furthermore, the basin suffered from more droughts during spring and summer than the other two seasons.

Major drought events

The spatial pattern and temporal evolution characteristics of the top five drought events were analyzed to further reveal the characteristics of severe drought events. The spatial pattern of the accumulated SEDI and the corresponding temporal evolution of the monthly characteristics (i.e. MS, MI, and MA) are shown in Figure 7. The track of the most severe drought (December 1962–October 1963) stretched across the entire basin. It emerged in the western part of the basin, then moved eastward, and finally wandered in the central part for 7 months until October 1963. The monthly affected area during the event varied considerably; the largest area

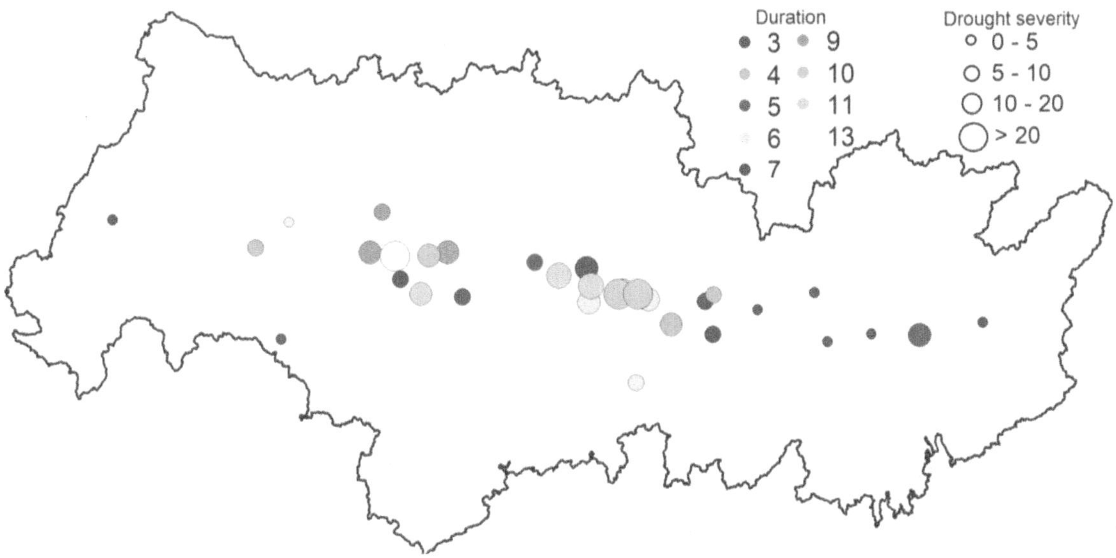

Figure 5 | Drought centroids, durations (month), and severities (10^5 km^2/month) during 1960–2015.

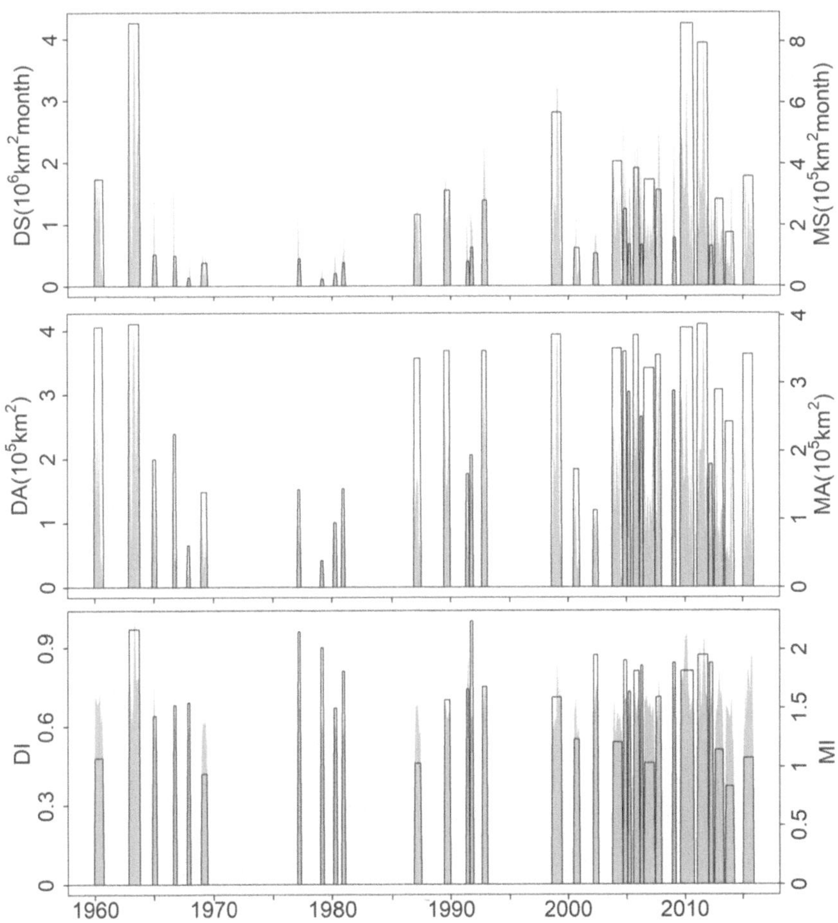

Figure 6 | Temporal variations of drought severity, affected area, and intensity for the total lifetime (i.e. DS, DA, and DI) and each monthly step (i.e. MS, MA, and MI). The black lines represent DS, DA, and DI, and the MS, MA, and MI are shown by the orange pillars. Please refer to the online version of this paper to see this figure in color: http://dx.doi.org/10.2166/nh.2021.061.

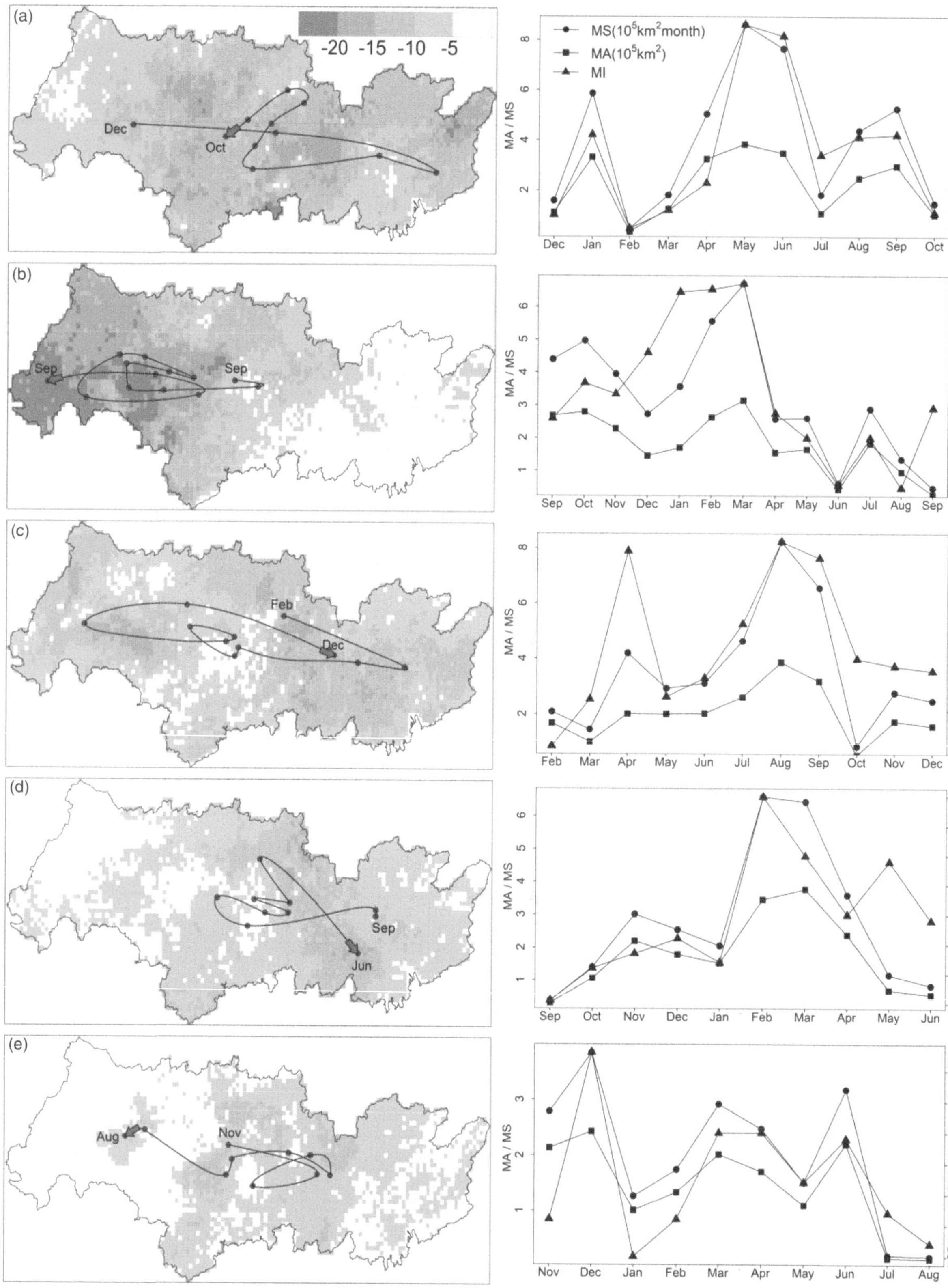

Figure 7 | Spatial distribution of the cumulated SEDI and temporal evolution of monthly characteristics for the top five drought events; (a–e) represent the drought periods of December 1962-October 1963, September 2009–September 2010, February 2011–December 2011, September 1998–June 1999, and November 2003–August 2004, respectively. The black dots indicate centroids of the drought cluster in each month, and the black lines between centroids reveal the drought paths with red arrows pointing to the destination.

affected occurred in May 1963, reaching up to 3.84×10^5 km^2 accounting for 84.6% of the basin. The evolutions of severity and intensity were also complex with multiple peaks. The drought from September 2009 to September 2010 was mainly concentrated on the eastern part of the basin with a circular track. It originated in the central part, wandered into the western part for 11 months, and disappeared in the westernmost part. The affected area suffered from extremely dry conditions from April to June 2010 with the largest value in June 2010, i.e. 3.15×10^5 km^2. For the February 2011–December 2011 drought event, it spread throughout the study region with an ambiguous track (Figure 7(c)). The drought originated in the mid-northern part, shifting to the eastern part afterwards and back to the central part later on, and then moved to the western part, ending in the eastern part. Although it lasted for 11 months with the largest monthly drought area of 3.87×10^5 km^2, only one peak of drought severity and intensity was detected. As for the 4th drought event, it was mainly concentrated in the central part, originating from the eastern part and moving into the central region. The drought lasted for 10 months from September 1998 to June 1999. According to the drought centroid, the pattern of the last drought event consisted of two sub-patterns, i.e. the central region (from November 2003 to June 2004) and the western region (July and August 2004). The drought path wandered in the central region and then moved westward. The monthly drought severity and area reached the maximum in the second month from its onset.

Drought impacts on NPP

The spatial distribution of correlation coefficients between NPP and SEDI during the four seasons of 1982–2015 was analyzed, and the results are shown in Figure 8. Overall, most of the basin was dominated by weak correlations in spring and summer, while significant positive correlations were detected in most of the basin during winter. According to the NPP cluster, the ratios in percentages between the grids with negative ΔNPP and total grids for 1982–2015 were calculated and are shown in Figure 9. In general, more than 50% of the drought region suffered from NPP losses during the 1982–2015 drought events. A loss of NPP was detected for nearly the entire drought region during two events (October 2004–January 2005 and February 2011–December 2011). For the total NPP loss, notably, the value was the largest during the February 2011–December 2011 drought, which reached 31.85 Tg C accounting for 11.7% of the regional annual mean NPP. The drought from September 2009 to September 2010 also caused large NPP loss up to 20.26 Tg C, accounting for 7.4% of the annual mean. Figure 10 further shows the temporal variations of drought severity, intensity, affected area, and total NPP loss during the drought events from 1982 to 2015. The total NPP loss showed significant correlations ($P < 0.05$) with drought severity and intensity, with correlation coefficients of 0.72 and 0.42, respectively.

Subsequently, the top two drought events with large NPP losses were selected to analyze the variations in NPP under severe drought events. The spatial pattern of cumulative NPP and the temporal evolution of monthly cumulative NPP, along with the correlations between the cumulative SEDI and NPP during the two drought events, are shown in Figure 11. A significant correlation ($P < 0.05$) was detected between cumulative SEDI and NPP for the two drought events. During the February 2011–December 2011 drought, a decrease in NPP was detected across nearly the entire basin. NPP showed a large

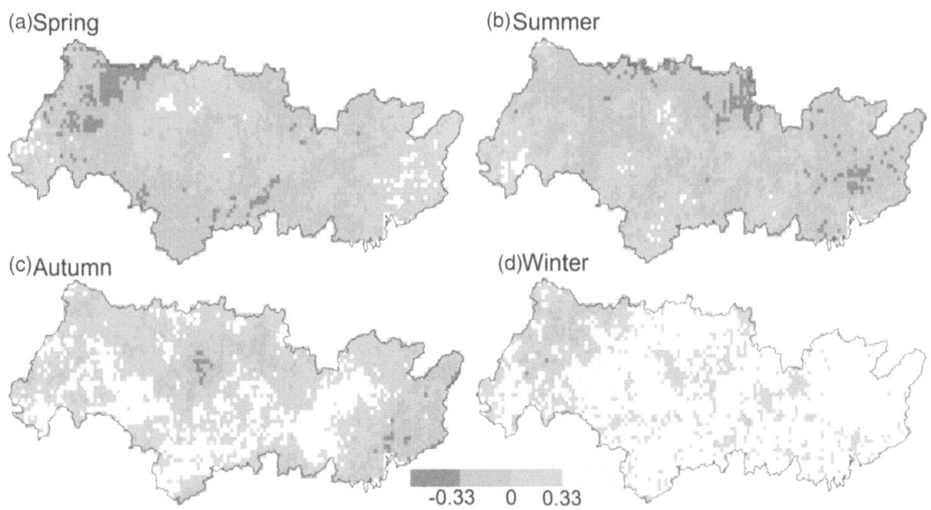

Figure 8 | Spatial distribution of correlation coefficient between SEDI and normalized NPP at four seasons from 1982 to 2015. Values larger than 0.33 or smaller than −0.33 indicate statistically significant correlations.

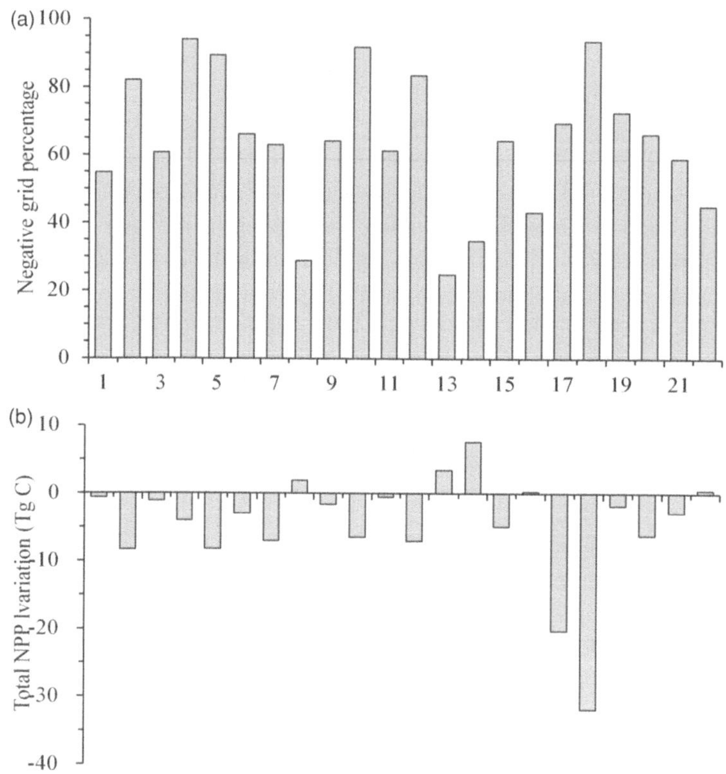

Figure 9 | (a) The percentage of grids with negative ΔNPP during the drought events arranged in chronological order for 1982–2015. (b) The total NPP variation during the drought events.

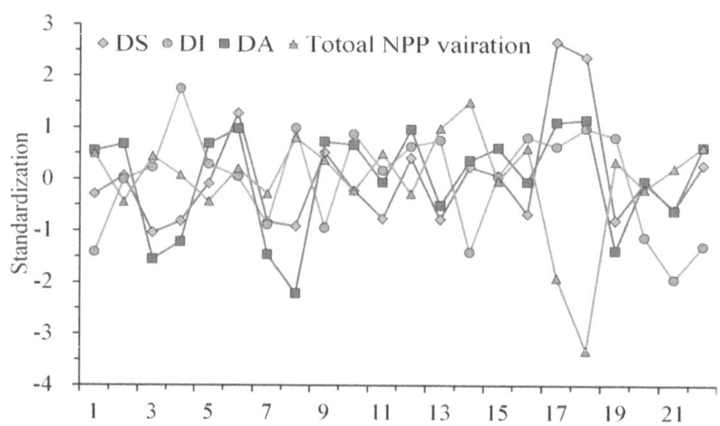

Figure 10 | Variation of normalized drought severity, intensity, affected area, and NPP variation during the drought events from 1982 to 2015.

reduction across the region with low SEDI, indicating that persistent drought severely affected NPP. The temporal variation of monthly cumulative NPP was generally consistent with monthly drought severity evolution; the NPP loss increased with higher drought severity. The September 2009–September 2010 drought caused a large decrease in NPP over the western part of the basin.

DISCUSSION

Variations in drought are affected by multiple factors, such as climate change, atmospheric circulation, and local geological conditions. Climate change has profound impacts on drought variation by changing climatic water supply and demand. A

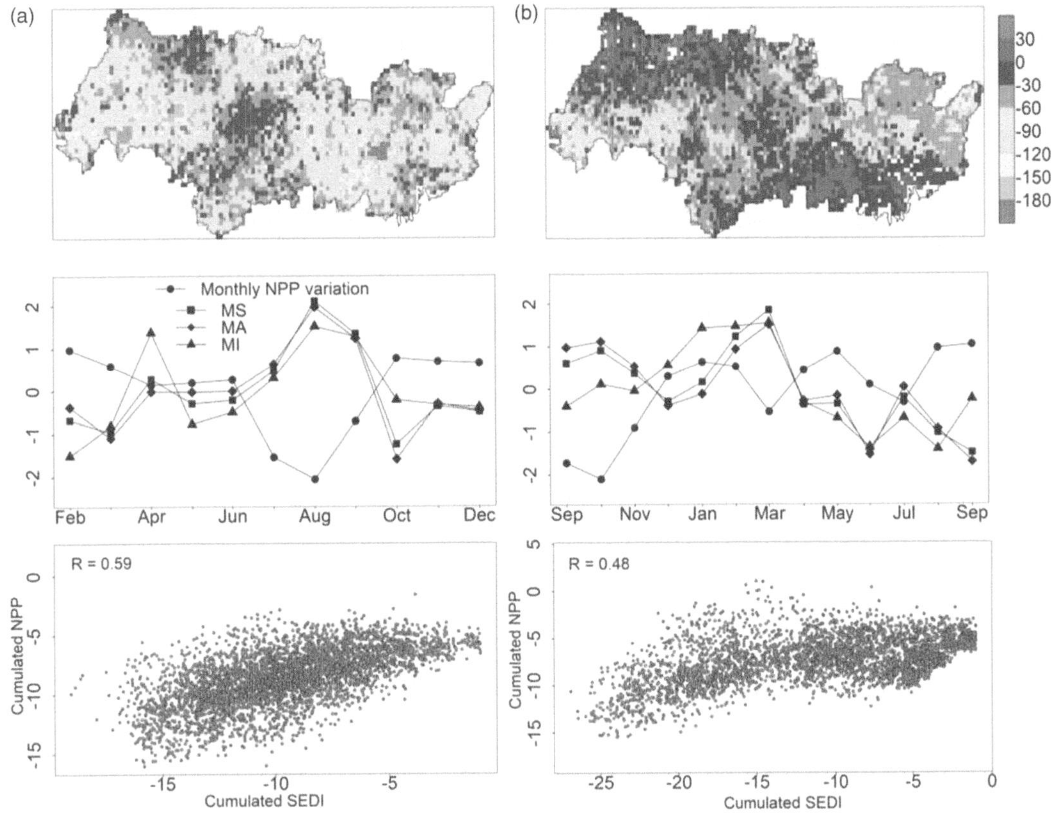

Figure 11 | Spatial distribution of cumulated NPP anomaly, temporal evolution of monthly drought characteristics (normalized MS, MI, and MA), and scatterplot between cumulated NPP and SEDI for the drought period of (a) February 2011–December 2011 and (b) September 2009–September 2010.

trend analysis of monthly SEDI from 1960 to 2015 was carried out to reveal the change in drought; variations in monthly precipitation, PET, and ETa were also analyzed, and the results are shown in Figure 12. A significant drying trend ($P < 0.05$) was found across most of the basins particularly for the upper reach. Precipitation tended to decrease, while PET and ETa generally increased significantly in the upper reach, which may largely contribute to the drying trend. Also, the droughts in the upstream may be partly affected by local complex topography. The upstream is characterized by karst topography which can trigger severe droughts (Guo *et al.* 2013). Hydrological characteristics over the karst region are found eventually altered by rocky desertification, particularly the rainfall–runoff process (Zhou 2020a). Rocky desertification features with little vegetation and soil breakage increase infiltration to subsurface systems reducing flow storage capacity and residence time and decreasing the resistance to overland flow, causing frequent drought and flood events in karst regions (Jiang *et al.* 2014; Yan & Cai 2015). Additionally, rainfall erosivity in some regions also shows an upward trend that the probability of water and soil erosion is increasing correspondingly (Peng & Wang 2012; Wang *et al.* 2013), which further exacerbates drought conditions in the upstream. On the other hand, because the basin is located in the monsoon region, the climate is strongly influenced by global circulation patterns, such as the El Nino-Southern Oscillation (ENSO) (Deng *et al.* 2018; Zhou *et al.* 2021). Large-scale atmospheric circulation can induce drought changes, as it alters the amount of water vapor available for precipitation and could affect the key components of the hydrologic cycle, such as soil moisture and ETa (Liu *et al.* 2013). The severe 2009–2010 drought over the upper reach was largely attributed to the concurrent distinctive warm pool El Niño that generated a strong anomalous cyclone over the northwestern Pacific and led to a serious reduction in rainfall over this region (Zhang *et al.* 2013). A study by Niu (2013) also shows a significant correlation between ENSO and precipitation variation in the eastern part of the Pearl River basin. By affecting precipitation patterns, the drought duration exhibited different patterns in different parts of the basin during warm/cool ENSO phases (Xiao *et al.* 2016).

We found that the most persistent drought events have occurred over the last decades in the basin. These drought events have visible impacts on agriculture, water resources, and ecosystems. The 2009–2010 drought with a duration of 13 months

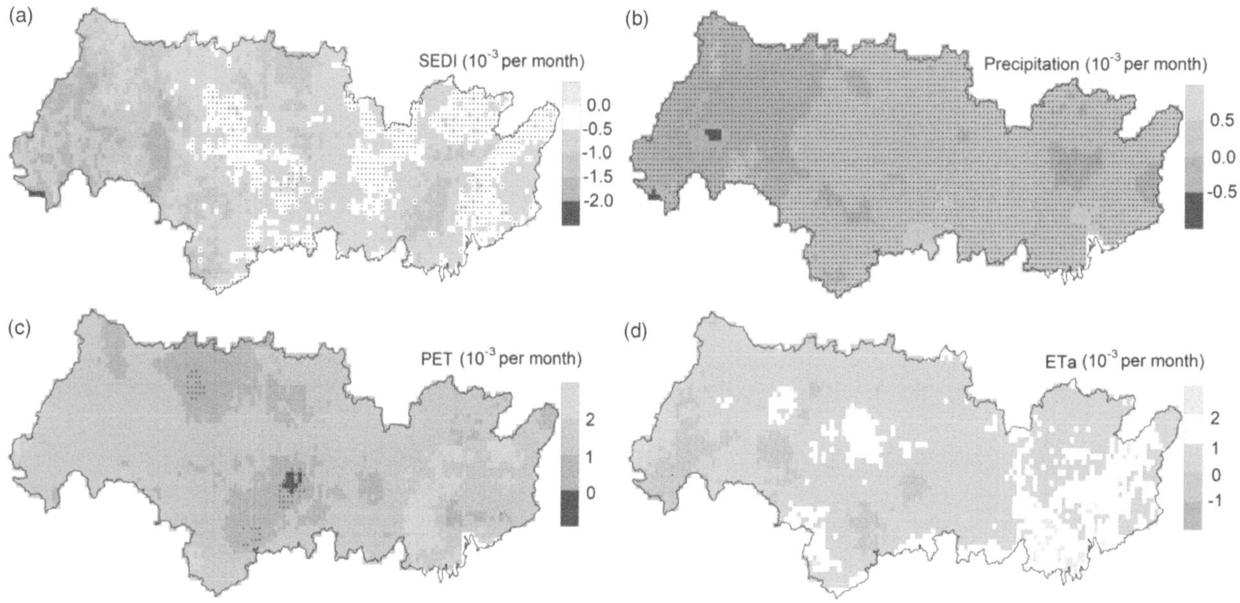

Figure 12 | Spatial pattern of trends in SEDI (a), precipitation (b), PET (c), and ETa (d) from 1960 to 2015. Regions not dotted indicate significant trends at 0.05 significance level.

nearly affected the entire basin, particularly the upper reach (Guangxi and south of Guizhou Province). It had large and destructive effects on agricultural production and the drinking water supply to residents (Barriopedro *et al.* 2012). During the drought period, numerous small and medium-sized rivers and reservoirs dried up in Guangxi and Guizhou (Lin *et al.* 2015). The drought has damaged winter wheat, causing a 30% reduction in production resulting in a temporary impact on local grain supply-demand balance (Li *et al.* 2010; Zhang *et al.* 2013). The freshwater supply of the Pearl River Delta, which is one of the most highly urbanized regions in the world, is facing an increasing risk in recent decades (Liu *et al.* 2014). The East River is one of the most important water supply resources for major cities in the Pearl River Delta such as Hong Kong, Guangzhou, Shenzhen, Dongguan, and Huizhou. However, the water supply is at an increasing risk because it suffers from more frequent dry seasons along with serious saltwater intrusion (Liu *et al.* 2016). To alleviate water supply pressure and ensure water security, water was delivered from the West River to the Pearl River Delta in recent years. Therefore, the increasingly frequent and severe drought events have caused far-reaching impacts on regional development, and more attention should be paid to drought risk over the basin in future.

Our study demonstrates that prolonged and severe drought events largely decreased regional NPP. For example, the 2009–2010 drought reduced NPP by 31.85 Tg C across the whole basin, accounting for 11.7% of the regional annual mean NPP. The occurrence of long-lasting and severe droughts could cause serious disturbances in the biochemical and physiological processes of ecosystems such as photosynthesis, respiration, and nitrogen and protein metabolism (Kurz *et al.* 2008; Zhang *et al.* 2012; Anderegg *et al.* 2013). Long-term dry conditions also influence ecosystem productivity by increasing pest and disease infestation. It is therefore necessary to manage water resources at the long-term time scale, such as monthly and seasonal reservoir storage management and water allocation (Zhou 2020b). While a prolonged and severe drought can lead to increasingly large NPP losses, drought events with short durations and low severities generally have less impact on NPP. Generally, the resilience and restorability of vegetation to drought stress is strong in humid regions. Most of the vegetation in humid regions possesses residual water and deep root systems that reduce the impact of short-term water shortages on plant productivity. In addition, the increase in radiation during drought periods can cause a decline in cloud cover and an increase in plant productivity (Saleska *et al.* 2007; Wang *et al.* 2015). These may explain the relatively lower NPP loss during moderate drought events in the basin.

CONCLUSIONS

In this study, the Pearl River basin was taken as the case region where severe drought events and their characteristics from 1960 to 2015 were analyzed using SEDI and the three-dimensional clustering algorithm. In addition, the impact of severe droughts on NPP was evaluated. It is shown that SEDI has the potential to identify drought events, and the basin experienced frequent extensive and severe drought events during the past two decades. About half of the drought events persisted for 6 months or longer, and some lasted for more than 9 months, which generally occurred in the upper and middle reaches. Results also indicate that drought events generally caused a reduction in NPP for more than half of the drought-affected area. The total NPP loss significantly correlated with drought severity and intensity. Typical examples further show that the total loss of NPP during the 2011 drought has reached 31.85 Tg C accounting for 11.7% of the annual mean, while the drought from September 2009 to September 2010 caused an NPP loss of 20.26 Tg C accounting for 7.4% of the annual mean.

ACKNOWLEDGEMENT

This research is supported by the National Natural Science Foundation of China (No. 51779067) and the Water Resource Science and Technology Innovation Program of Guangdong Province (2020–29). The authors thank Drs Zhaoli Wang and Xushu Wu at South China University of Technology for technical help.

CONFLICT OF INTEREST

The authors declare that there are no conflicts of interest.

DATA AVAILABILITY STATEMENT

All relevant data are included in the paper or its Supplementary Information.

REFERENCES

Anderegg, W. R., Kane, J. M. & Anderegg, L. D. 2013 Consequences of widespread tree mortality triggered by drought and temperature stress. *Nat. Clim. Change* **3** (1), 30.

Anderegg, W. R., Flint, A., Huang, C. Y., Flint, L. & Berry, J. A. 2015 Tree mortality predicted from drought-induced vascular damage. *Nat. Geosci.* **8** (5), 367.

Ayantobo, O. O., Li, Y., Song, S. & Yao, N. 2017 Spatial comparability of drought characteristics and related return periods in mainland China over 1961–2013. *J. Hydrol.* **550**, 549–567.

Barriopedro, D., Gouveia, C. M., Trigo, R. M. & Wang, L. 2012 The 2009/10 drought in China: possible causes and impacts on vegetation. *J. Hydrometeorol.* **13** (4), 1251–1267.

Belayneh, A., Adamowski, J., Khalil, B. & Ozga-Zielinski, B. 2014 Long-term SPI drought forecasting in the Awash River Basin in Ethiopia using wavelet neural network and wavelet support vector regression models. *J. Hydrol.* **508**, 418–429.

Castro, S. M., Sanchez-Azofeifa, G. A. & Sato, H. 2018 Effect of drought on productivity in a Costa Rican tropical dry forest. *Environ. Res. Lett.* **13**, 4.

Chen, T., Werf, G. R., Jeu, R. A. M., Wang, G. & Dolman, A. J. 2013 A global analysis of the impact of drought on net primary productivity. *Hydrol. Earth Sys. Sci.* **17**, 3885–3894.

Deng, S., Chen, T., Yang, N. & Qu, L. 2018 Spatial and temporal distribution of rainfall and drought characteristics across the Pearl River basin. *Sci. Total Environ.* **619**, 28–41.

Gocic, M. & Trajkovic, S. 2014 Spatiotemporal characteristics of drought in Serbia. *J. Hydrol.* **510**, 110–123.

Guo, F., Jiang, G., Yuan, D. & Polk, J. S. 2013 Evolution of major environmental geological problems in karst areas of Southwestern China. *Environ. Earth Sci.* **69** (7), 2427–2435.

Guo, H., Bao, A., Ndayisaba, F., Liu, T. & Jiapaer, G. 2018 Space-time characterization of drought events and their impacts on vegetation in Central Asia. *J. Hydrol.* **564**, 1165–1178.

Herrera-Estrada, J. E., Satoh, Y. & Sheffield, J. 2017 Spatiotemporal dynamics of global drought. *Geophys. Res. Lett.* **44** (5), 2254–2263.

Jiang, Z., Lian, Y. & Qin, X. 2014 Rocky desertification in Southwest China: impacts, causes, and restoration. *Earth Sci. Rev.* **132**, 1–12.

Kao, I.-F., Zhou, Y., Chang, L.-C. & Chang, F.-J. 2020 Exploring a long short-term memory based encoder-decoder framework for multi-step-ahead flood forecasting. *J. Hydrol.* **583**, 124631.

Kurz, W. A., Stinson, G., Rampley, G. J. & Dymond, C. C. 2008 Risk of natural disturbances makes future contribution of Canada's forests to the global carbon cycle highly uncertain. *Proc. Natl. Acad. Sci. U. S. A.* **105** (5), 1551–1555.

Lai, C., Li, J., Wang, Z., Wu, X., Zeng, Z., Chen, X., Lian, Y., Yu, H., Wang, P. & Bai, X. 2018 Drought-induced reduction in net primary productivity across mainland China from 1982 to 2015. *Remote Sens.* **10**, 1433.

Lesk, C., Rowhani, P. & Ramankutty, N. 2016 Influence of extreme weather disasters on global crop production. *Nature* **529** (7584), 84.

Li, Q., Yan, N., Zhang, F., Chang, S. & Wu, B. 2010 Drought monitoring and its impacts assessment in southwest China using remote sensing in the spring of 2010. *Acta Geogr. Sin.* **65**, 771–780.

Li, J., Wang, Z., Wu, X., Xu, C.-Y., Guo, S. & Chen, X. 2020a Toward monitoring short-term droughts using a novel daily scale, standardized antecedent precipitation evapotranspiration index. *J. Hydrometeorol.* **21**, 891–908.

Li, J., Wang, Z., Wu, X., Chen, J., Guo, S. & Zhang, Z. 2020b A new framework for tracking flash drought events in space and time. *Catena* **194**, 104763.
Li, J., Wang, Z., Wu, X., Guo, S. & Chen, X. 2020c Flash droughts in the Pearl River basin, China: observed characteristics and future changes. *Sci. Total Environ.* **707**, 136074.
Li, J., Wang, Z., Wu, X., Zscheischler, J., Guo, S. & Chen, X. 2021a A standardized index for assessing sub-monthly compound dry and hot conditions with application in China. *Hydrol. Earth Syst. Sci.* **25**, 1587–1601.
Li, J., Wang, Z., Wu, X., Xu, C., Guo, S., Chen, X. & Zhang, Z. 2021b Robust meteorological drought prediction using antecedent SST fluctuations and machine learning. *Water Resour. Res.* **27**, e2020WR029413. https://doi.org/10.1029/2020WR029413.
Liang, W., Yang, Y., Fan, D., Guan, H. & Zhang, T. 2015 Analysis of spatial and temporal patterns of net primary production and their climate controls in China from 1982 to 2010. *Agr. Forest Meteorol.* **204**, 22–36.
Lin, W., Wen, C. & Wen, Z. 2015 Drought in Southwest China: a review. *Atmos. Oceanic Sci. Lett.* **8** (6), 339–344.
Lin, K., Sheng, S., Zhou, Y., Liu, F., Li, Z., Chen, H., Xu, C.-Y., Chen, J. & Guo, S. 2020 The exploration of a temporal convolutional network combined with encoder-decoder framework for runoff forecasting. *Hydrol. Res.* **51** (5), 1136–1149.
Liu, Z., Zhou, P., Zhang, F., Liu, X. & Chen, G. 2013 Spatiotemporal characteristics of dryness/wetness conditions across Qinghai Province, Northwest China. *Agr. Forest Meteorol.* **182**, 101–108.
Liu, D., Guo, S., Shao, Q., Jiang, Y. & Chen, X. 2014 Optimal allocation of water quantity and waste load in the Northwest Pearl River Delta, China. *Stoch. Environ. Res. Risk Assess.* **28** (6), 1525–1542.
Liu, B., Chen, J., Lu, W. & Chen, X. 2016 Spatiotemporal characteristics of precipitation changes in the Pearl River basin, China. *Theor. Appl. Climatol.* **123** (3–4), 537–550.
Liu, Y., Zhu, Y., Ren, L., Singh, V. P. & Yong, B. 2019 Understanding the spatiotemporal links between meteorological and hydrological droughts from a three-dimensional perspective. *J. Geophys. Res. Atmos.* **124** (6), 3090–3109.
Lloyd-Hughes, B. 2012 A spatio-temporal structure-based approach to drought characterization. *Int. J. Climatol.* **32** (3), 406–418.
Niu, J. 2013 Precipitation in the Pearl River basin, South China: scaling, regional patterns, and influence of large-scale climate anomalies. *Stoch. Environ. Res. Risk Assess.* **27** (5), 1253–1268.
Niu, J. & Chen, J. 2016 A wavelet perspective on variabilities of hydrological processes in conjunction with geomorphic analysis over the Pearl River basin in South China. *J. Hydrol.* **542**, 392–409.
Peng, T. & Wang, S. 2012 Effects of land use, land cover and rainfall regimes on the surface runoff and soil loss on karst slopes in southwest China. *Catena* **90**, 53–62.
Piao, S., Fang, J., Ciais, P., Peylin, P. & Huang, Y. 2009 The carbon balance of terrestrial ecosystems in China. *Nature* **458** (7241), 1009.
Saleska, S. R., Didan, K. & Huete, A. R. 2007 Amazon forests green-up during 2005 drought. *Science* **318** (5850), 612–612.
Sander, J., Ester, M., Kriegel, H. P. & Xu, X. 1998 Density-based clustering in spatial databases: the algorithm GDBSCAN and its applications. *Data Min. Knowl. Discov.* **2** (2), 169–194.
Saxton, K. E. & Rawls, W. J. 2006 Soil water characteristic estimates by texture and organic matter for hydrologic solutions. *Soil Sci. Soc. Am. J.* **70** (5), 1569–1578.
Sheffield, J., Andreadis, K. M., Wood, E. F. & Lettenmaier, D. P. 2009 Global and continental drought in the second half of the twentieth century: severity–area–duration analysis and temporal variability of large-scale events. *J. Climate* **22** (8), 1962–1981.
Su, B., Huang, J., Fischer, T., Wang, Y., Kundzewicz, Z. W. & Zhai, J. 2018 Drought losses in China might double between the 1.5° C and 2.0° C warming. *Proc. Natl. Acad. Sci. U. S. A.* **115** (42), 10600–10605.
Sun, B., Zhao, H. & Wang, X. 2016 Effects of drought on net primary productivity: roles of temperature, drought intensity, and duration. *Chinese Geogr. Sci.* **26** (2), 270–282.
Tijdeman, E., Stahl, K. & Tallaksen, L. M. 2020 Drought characteristics derived based on the standardized streamflow index: a large sample comparison for parametric and nonparametric methods. *Water Resour. Res.* **56** (10), e2019WR026315.
Vicente-Serrano, S. M., Camarero, J. J., Zabalza, J., Sangüesa-Barreda, G., López-Moreno, J. I. & Tague, C. L. 2015 The evapotranspiration deficit controls growth and net primary production: implications for Circum-Mediterranean forests under forecasted warmer and drier conditions. *Agr. Forest Meteorol.* **206**, 45–54.
Vicente-Serrano, S. M., Miralles, D. G., Domínguez-Castro, F., Azorin-Molina, C., Kenawy, A. & McVicar, T. R. 2018 Global assessment of the Standardized Evapotranspiration Deficit Index (SEDI) for drought analysis and monitoring. *J. Climate* **31** (14), 5371–5393.
Wang, B., Zheng, F., Römkens, M. J. & Darboux, F. 2013 Soil erodibility for water erosion: a perspective and Chinese experiences. *Geomorphology* **187**, 1–10.
Wang, H., Chen, A., Wang, Q. & He, B. 2015 Drought dynamics and impacts on vegetation in China from 1982 to 2011. *Ecol. Eng.* **75**, 303–307.
Wang, L., Yuan, X., Xie, Z., Wu, P. & Li, Y. 2016 Increasing flash droughts over China during the recent global warming hiatus. *Sci. Rep.* **6**, 30571.
Wu, H., Hayes, M. J., Weiss, A. & Hu, Q. 2001 An evaluation of the Standardized Precipitation Index, the China-Z Index and the statistical Z-Score. *Int. J. Climatol.* **21** (6), 745–758.
Wu, C., Huang, G., Yu, H., Chen, Z. & Ma, J. 2014 Impact of climate change on reservoir flood control in the upstream area of the Beijiang River Basin, South China. *J. Hydrometeorol.* **15** (6), 2203–2218.

Wu, X., Wang, Z., Zhou, X., Lai, C. & Chen, X. 2017a Trends in temperature extremes over nine integrated agricultural regions in China, 1961–2011. *Theor. Appl. Climatol.* **129**, 1279–1294.

Wu, X., Wang, Z., Guo, S., Liao, W., Zeng, Z. & Chen, X. 2017b Scenario-based projections of future urban inundation within a coupled hydrodynamic model framework: a case study in Dongguan City, China. *J. Hydrol.* **547**, 428–442.

Wu, X., Guo, S., Yin, J., Yang, G., Zhong, Y. & Liu, D. 2018 On the event-based extreme precipitation across China: time distribution patterns, trends, and return levels. *J. Hydrol.* **562**, 305–317.

Xia, Y., Sheffield, J., Ek, M. B., Dong, J., Chaney, N. & Wei, H. 2014 Evaluation of multi-model simulated soil moisture in NLDAS-2. *J. Hydrol.* **512**, 107–125.

Xiao, M., Zhang, Q., Singh, V. P. & Liu, L. 2016 Transitional properties of droughts and related impacts of climate indices in the Pearl River basin, China. *J. Hydrol.* **534**, 397–406.

Xu, K., Yang, D., Yang, H., Li, Z. & Qin, Y. 2015 Spatio-temporal variation of drought in China during 1961–2012: a climatic perspective. *J. Hydrol.* **526**, 253–264.

Yan, X. & Cai, Y. 2015 Multi-scale anthropogenic driving forces of karst rocky desertification in Southwest China. *Land Degrad. Dev.* **26** (2), 193–200.

Zhang, L., Xiao, J., Li, J., Wang, K., Lei, L. & Guo, H. 2012 The 2010 spring drought reduced primary productivity in southwestern China. *Environ. Res. Lett.* **7** (4), 045706.

Zhang, W., Jin, F., Zhao, J. & Qi, L. 2013 The possible influence of a nonconventional El Niño on the severe autumn drought of 2009 in Southwest China. *J. Climate* **26** (21), 8392–8405.

Zhang, L., Qin, X., Liu, P., Huang, Q. & Lan, F. 2015 Estimation of carbon sink fluxes in the Pearl River basin (China) based on a water–rock–gas–organism interaction model. *Environ. Earth Sci.* **74** (2), 945–952.

Zhou, Y. 2020a Real-time probabilistic forecasting of river water quality under data missing situation: deep learning plus post-processing techniques. *J. Hydrol.* **598**, 125164.

Zhou, Y. 2020b Exploring multidecadal changes in climate and reservoir storage for assessing nonstationarity in flood peaks and risks worldwide by an integrated frequency analysis approach. *Water Res.* **185**, 116265.

Zhao, M. & Running, S. W. 2010 Drought-induced reduction in global terrestrial net primary production from 2000 through 2009. *Science* **329** (5994), 940–943.

Zhou, Y., Guo, S., Xu, C.-Y., Chen, H., Guo, J. & Lin, K. 2016 Probabilistic prediction in ungauged basins (PUB) based on regional parameter estimation and Bayesian model averaging. *Hydrol. Res.* **47** (6), 1087–1103.

Zhou, Y., Guo, S. & Chang, F.-J. 2019 Explore an evolutionary recurrent ANFIS for modelling multi-step-ahead flood forecasts. *J. Hydrol.* **570**, 343–355.

Zhou, Y., Guo, S., Xu, C-Y., Chang, F. & Yin, J. 2020 Improving the reliability of probabilistic multi-step-ahead flood forecasting by fusing unscented Kalman filter with recurrent neural network. *Water* **12** (2), 578.

Zhou, Y., Zuo, Y., Zhang, Y., Jin, J., Zhou, P., Wu, C., Cui, Y. & Ning, S. 2021 Identification and characteristics analysis of Meiyu in Anhui Province based on the National Standard of Meiyu monitoring indices. *Hydrol. Res.* In press, doi: 10.2166/nh.2021.042.

First received 20 June 2021; accepted in revised form 30 July 2021. Available online 17 August 2021

doi: 10.2166/nh.2021.178

Copula-based modeling of hydraulic structures using a nonlinear reservoir model

Qiaofeng Tan, Yunze Mao, Xin Wen*, Tian Jin, Ziyu Ding and Zhenni Wang

College of Water Conservancy and Hydropower Engineering, Hohai University, Nanjing 210098, China
*Corresponding author. E-mail: njwenxin@163.com

YM, 0000-0001-6611-7074

ABSTRACT

Multivariate flood frequency analysis has been widely used in the design and risk assessment of hydraulic structures. However, analytical solutions are often obtained based on an idealized linear reservoir model in which a linear routing process is assumed, and consequently, the flood risk is likely to be over- or underestimated. The present study proposes a nonlinear reservoir model in which the relationships of reservoir water level with reservoir volume and discharge are assumed to be nonlinear in order to more accurately describe the routing process as it takes into consideration the interactions between hydrological loading and different discharge structures. The structure return period is calculated based on the copula function and compared with that based on the linear reservoir model and the bivariate return period based on the Kendall distribution function. The results show that the structure return period based on the linear model leads to an underestimation of the flood risk under the conditions of high reservoir water level. For the same reservoir, linear and nonlinear reservoir models give quite different reservoir volume-water level and discharge-water level curves; therefore, they differ substantially in the sensitivity to flood events with different combinations of flood peak and volume. We also analyze the effects of the parameters involved in the reservoir volume-water level and discharge-water level relationships on the maximum water level at different return periods in order to better understand the applicability and effectiveness of the proposed method for different hydraulic projects.

Key words: bivariate flood frequency analysis, copula, nonlinear reservoir model, structure return period

HIGHLIGHTS

- The structure return period is calculated based on the nonlinear reservoir model.
- Nonlinear relationships of water level with volume and discharge are assumed.
- Interactions between hydrological loading and different structures are analyzed.
- Effects of different storage and discharge capacity parameters are discussed.

1. INTRODUCTION

An extreme hydrological event is considered a major cause of hydraulic structure failure, and thus, its occurrence probability provides an important basis for the evaluation of the structure return period in the design of hydraulic structures. Thus, it is important to choose appropriate variables and methods for the risk assessment of an extreme hydrological event acting on the hydraulic structures. However, the widely used univariate flood frequency analysis does not account for the interactions between hydrological events and hydraulic structures. The extreme load that can be supported by a hydraulic structure in a flood event is determined by the storage and routing processes, and changes in any characteristic variable of the flood event, such as flood peak, volume and duration, and the hydraulic structure, can have a substantial effect on the extreme load.

In univariate flood frequency analysis, the return period of a given hydrological variable is taken as the return period of structural failure. It is evident that the univariate return period of the flood peak or volume cannot provide a complete evaluation of the occurrence probability of hydrological events (Chebana & Ouarda 2011), which therefore may result in over- or underestimation of the flood risk and hydraulic design standard (Salvadori & De Michele 2004). Apparently, multivariate flood frequency analysis is preferred over univariate flood frequency analysis in hydraulic design. However, a number of drawbacks have also been identified in previous literature. For instance, a normal distribution is often assumed for non-normally distributed flood peak and volume in multivariate normal distribution and normal transformation methods.

Some marginal distributions may be combined to form a joint distribution, such as bivariate lognormal distribution, bivariate exponential distribution, bivariate Gumbel distribution, and two-dimensional P-III distribution. However, it is generally assumed that the marginal distributions should come from the same family of distributions.

Recently, the Copula function has been widely used in hydrology, especially in multivariate flood frequency analysis. The copula is a function that links univariate distribution functions to a multivariate distribution function describing the dependence among correlated variables (Nelsen 2006). However, there is no universally agreed method for calculating the multivariate return period based on the copula function. Salvadori & De Michele (2004) investigated both unconditional and conditional return periods of hydrological events using copulas, and they found that it was possible to combine the marginal events in several ways using the (inclusive) OR operator and the AND operator. Salvadori *et al.* (2011) introduced another return period, the secondary return period (also called the Kendall return period), which was associated with the realization of dangerous events for the dam. It should be noted that the return periods mentioned above are defined based on the natural occurrence probability of floods. However, the climate changes and human activities in recent years may make the hydrological extreme value series highly non-stationary. For the frequency analysis of non-stationary hydrological extreme value series in a changing environment, the return period should be redefined and several non-stationary methods have been developed, such as expected waiting time, expected number of exceedances, equivalent reliability, and average design life level (Yan *et al.* 2017, 2020).

However, the return period of hydraulic structures should be defined in terms of the risk of dam overtopping or the downstream loss. In multivariate cases, structural failure is considered to be the result of the complex interactions between the structure and the hydrological loads acting on the structure, which indicates that the effects of the reservoir and dam on the inflow hydrographs should also be taken into consideration (Mediero *et al.* 2010). The relationships between hydrological and structure variables can be simulated using observed data. De Michele *et al.* (2005) generated peak-volume pairs using Gumbel copula in order to verify that the maximum water level reached at the dam by the generated hydrographs was below the crest level. Klein *et al.* (2010) proposed a methodology to categorize flood events for risk-based analysis and the design of flood control systems using copulas, and the maximum water level reached at the dam was estimated to graphically analyze its relation with the primary return period for each event. Requena *et al.* (2013) compared the multivariate return period of the hydrological loads with the structure return period by simulation.

An idealized reservoir model can also be established to simulate the relationships between hydrological and structure variables. Notably, the reservoir volume-water level and discharge-water level relationships are assumed to be linear due to the inherent difficulties in mathematical derivation. For instance, Volpi & Flori (2014) proposed an idealized reservoir model in which the reservoir volume-water level and discharge-water level relationships were assumed to be linear and the inflow process was assumed to be constant, based on which the relationships of the maximum discharge with flood peak and volume were derived. Balistrocchi *et al.* (2017) proposed triangular hydrographs by assuming linear reservoir volume-water level and discharge-water level relationships. However, the reservoir volume-water level and discharge-water level relationships are more likely to be nonlinear rather than linear in actual reservoir operation depending on the hydraulic structures such as sluice gates, overflow weirs, and orifices. Thus, linearization can lead to over- or under-estimation of the structure return period and flood risk.

The main purpose of the present study is to determine the relationships of the maximum water level at the dam with flood peak and volume by assuming nonlinear reservoir volume-water level and discharge-water level relationships and to develop a methodology for calculating the structure return period based on the copula function. The rest of this paper is organized as follows. In Section 2, we illustrate the structure-based nonlinear multivariate approach for hydraulic structure design and risk assessment. For simplicity, only the bivariate case is considered here. In Section 3, we develop an idealized nonlinear reservoir model, based on which the calculation method of the structure return period is derived. In Section 4, the structure return period based on the nonlinear reservoir model is calculated and then compared with that based on the linear reservoir model and the bivariate return period based on the Kendall distribution function. In Section 5, we discuss the effects of the discharge and storage capacity of the reservoir on the return period. Finally, the main conclusions are drawn in Section 6.

2. DEFINITION OF RETURN PERIOD

2.1. Return period

The return period is used in hydrology to estimate the recurrence interval of a hydrological event such as flood, rainstorm, and extremely high or low water level with the severity or duration greater than or equal to a given value.

In univariate analysis, X is a random variable that measures the hydrological loads such as flood peak or volume, and F_X is the probability distribution function of X. The occurrence of X is assumed to be time-independent. The return period of $\{X > x\}$ is expressed as $T(x) = \mu/[1 - F_X(x)]$, where μ is the average time interval between two successive occurrences of the hydrological event X ($\mu = 1$ year in the case of annual maxima).

In multivariate analysis, some or several hydrological parameters are taken into consideration. For simplicity, only the bivariate case is considered in this study. Let X and Y be two random hydrological variables that measure the hydrological processes acting on the structure (e.g., flood peak and volume), and $F_{X,Y}$ be the joint cumulative probability distribution function of X and Y:

$$F_{X,Y}(x, y) = P_r[(X < x) \cap (Y < y)] \tag{1}$$

The joint probability density function $f_{X,Y}$ can be described as follows:

$$f_{X,Y}(x, y) = \frac{\partial^2 F_{X,Y}(x, y)}{\partial x \partial y} \tag{2}$$

provided that $f_{X,Y}$ does exist.

2.2. Estimation of the bivariate return period by copulas

Copula has emerged as a useful tool for multivariate hydrological frequency analysis, and the Archimedean and extreme value copula families are most often used for modeling flood variables. A distinct property of Archimedean copulas is that they can be constructed easily and allow for a great variety of dependence structures. Archimedean copulas such as Frank copula (Favre *et al.* 2004) and Clayton copula (Shiau *et al.* 2006) have been widely used to characterize the dependence structure between peak and volume variables. Extreme value copulas are able to connect the extreme values of the studied variables, which are very important in flood frequency analysis. The Gumbel copula is considered to best represent the relationship between peak and volume (Zhang & Singh 2006). Many bivariate return periods have been estimated by copulas in recent years, and the most representative ones are the 'AND' return period and the Kendall return period.

For a given critical probability level p, the 'AND' return period is described as:

$$\theta = \frac{\mu}{1 - p} = \frac{\mu}{1 - F_{X,Y}(x_p, y_p)} \tag{3}$$

where x_p and y_p are the coordinate pairs of any point on the critical level curve L_p.

The critical level curve L_p is the set of points corresponding to the probability level p, $L_p = \{(x, y) \in R^2 | F_{X,Y}(x, y) = p\}$, and the corresponding supercritical region is given by $\widehat{D}_p = \{(x, y) \in R^2 | (x > x_p) \cup (y > y_p)\}$, in which X and Y can be infinite, and the coordinate pairs of x_p and y_p on the critical level curve L_p represent the probability p.

The Kendall return period is defined in order to identify the univariate critical threshold in a multivariate context (Salvadori *et al.* 2013), and it is described as:

$$\theta_K = \frac{\mu}{1 - K(p)} \tag{4}$$

where $K(p)$ is the Kendall distribution function obtained from the joint distribution function $F_{X,Y}$, which is expressed as $K(p) = \Pr[F_{X,Y}(x, y) \leq p]$. $K(p)$ is greater than or equal to K, and then $\theta_K \geq \theta$. This return period represents the average inter-arrival time of random variables X and Y in the supercritical region $\widehat{D}_K$, $\widehat{D}_K = \{(x, y) \in R^2 | F_{X,Y}(x, y) > p\}$. The critical level curve can be further improved based on the 'AND' return period, which is thus referred to as the secondary return period.

2.3. Structure return period

The concept of structure return period is inspired by the supercritical region of the Kendall return period. It is noted that both 'AND' and Kendall return periods account only for the natural occurrence of floods, while the structure return period takes into consideration the response of hydraulic structures of the reservoir to floods. In this case, the critical level depends critically on the consistent responses of the hydraulic structures, and the supercritical region is the region over the structure response isoline.

Z is defined as the design parameter characterizing a hydraulic structure (e.g., the dimension of the spillway and the elevation of the levee) or variables associated with the structure. In actual applications, Z indicates the effect of hydrological loads acting on the structure. There is a strictly monotonically increasing function between Z and hydrological load X in a univariate case. The structure function is $Z = g(x)$, where there exists a unique inverse function g^{-1}. Note that this assumption holds for all hydraulic structures. In a univariate case, the probability distribution function of Z is the same as that of X, and thus $F_Z(z) = F_X(x)$, where $x = g^{-1}(z)$. Then, the return period of Z is the same as that of X, both of which can be used in the evaluation of hydrological events.

In a bivariate case, the structure design parameter Z depends on two hydrologic variables, X and Y, where $Z = g(X,Y)$, and the structure function $g(X,Y)$ takes into consideration the interactions of the structure and the hydrological load acting on the structure. The joint probability distribution function of Z can be derived from the following equation:

$$F_Z(z) = \int_{D_Z} f_{X,Y}(x, y)\, dxdy \tag{5}$$

where D_Z is the region of the XY plane that satisfies $g(x, y) \leq z$. In order to obtain $F_Z(z)$, there is a need to find the upper integral of D_Z for each z.

The return period of structural failure can be easily obtained by using the standard univariate frequency analysis of the random variable Z, $T(z) = \mu/[1 - F_Z(z)]$, which is the average time interval between two successive events in the same supercritical region $\widehat{D}_Z = \{(x, y) \in R^2 | g(x, y) > z\}$. A vector of the design parameter Z can also be obtained from the return period, and the expression of D_Z may be very complex in this case. Thus, only a single design variable Z is considered in this study for the sake of simplicity.

Then, the design parameter quantile in multivariate design can be determined:

$$z_T = F_Z^{-1}\left(\frac{1-\mu}{T}\right) \tag{6}$$

where F_Z can be determined by Equation (5).

Here, the objective of hydrological design is to determine the combination of x_p and y_p pairs in a given supercritical region at a given failure probability p. The supercritical regions may differ substantially in terms of their boundaries, which is closely associated with the structure under consideration and associated variables. The boundary of the supercritical region $\widehat{D}_Z$ can be determined by the structure function $g(x,y)$ described in Equation (5), which represents the hydrological load–structure interactions. Thus, it is imperative to consider the structure and the way it interacts with the hydrological load in determining the suitable structure function $g(x,y)$. However, it is noted that the expression of $g(x,y)$ is too complex to be described mathematically in many actual cases, and thus, there is a need to obtain the analytical solutions in a simpler and more efficient fashion.

3. NONLINEAR STRUCTURE MODELING

It is clear that the structure function $g(x,y)$ that can more accurately capture the interactions between the structure and the hydrological load acting on the structure plays a key role in calculating the structure return period. Given the complexity in mathematical derivation, an idealized reservoir model was proposed in previous studies of Volpi & Flori (2014) and Balistrocchi *et al.* (2017), where the reservoir volume-water level and discharge-water level relationships were assumed to be linear. However, both of them are likely to be nonlinear rather than linear in actual reservoir operation depending on the combinations of different drainage facilities such as sluice gates, overflow weirs, orifices and hydraulic turbines, as shown in Figure 1. Thus, this study focuses on the structure return period based on a nonlinear reservoir model. First, a nonlinear reservoir model is developed. Then, reservoir operation is simulated for different inflow hydrographs based on the water balance principle, and the relationships between the design variable (the maximum discharge) and two flood variables (flood peak and volume) are determined.

The nonlinear reservoir model is schematically shown in Figure 1. Ω is the reservoir area, which is assumed to increase continuously with the increase of water level H ($H_0 = 0$ m at the beginning of flood discharge), and the reservoir volume V and discharge Q_{out} are also calculated from $H_0 = 0$.

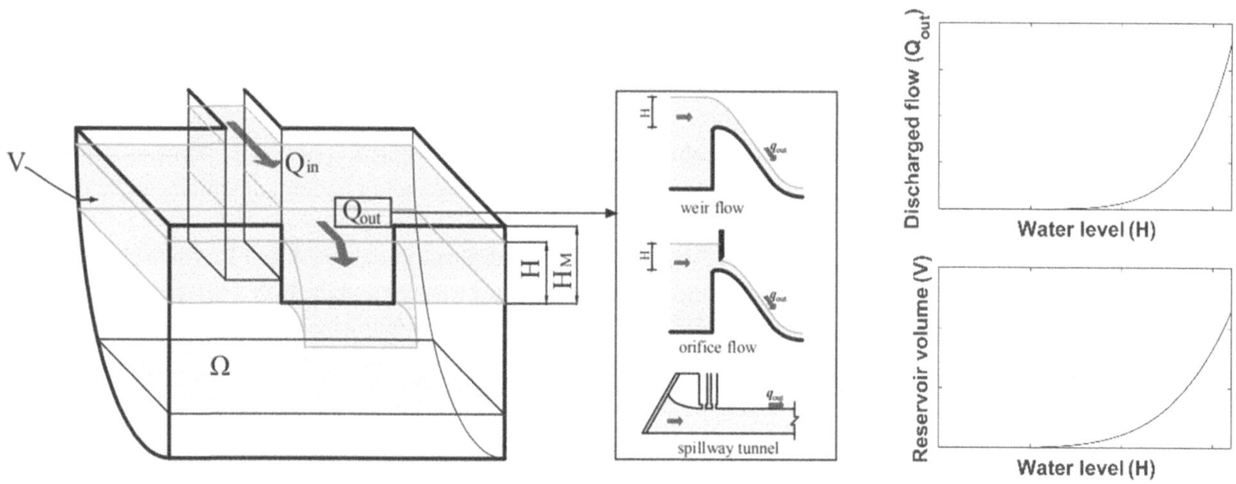

Figure 1 | The nonlinear reservoir model. Q_{in} is the inflow hydrograph; Ω is the reservoir area; V and H are the reservoir volume and water level above the spillway crest, respectively; H_M is the maximum water level during the routing process; and Q_{out} is the output hydrograph.

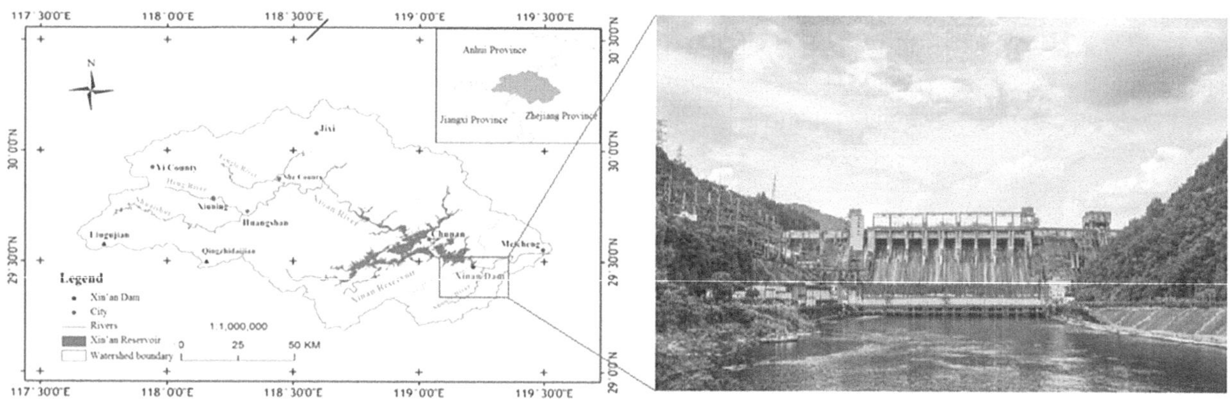

Figure 2 | The river networks of Xin'an basin and the location of Xin'an reservoir.

As the reservoir volume-water level and discharge-water level relationships are assumed to be nonlinear, the relationship between reservoir volume V and water level H can be described as follows:

$$V = K_1 H^{n/m} \tag{7}$$

where V and H are the reservoir volume and water level above the spillway crest, respectively.

The relationship between the reservoir discharge Q_{out} and the water level H is:

$$Q_{out} = K_2 H^{1/m} \tag{8}$$

where Q_{out} is the output hydrograph.

It is known that $1/m$ is associated with different discharge structures, which is often taken to be $1/m = 0.5$ for orifice outflow and $1/m = 1.5$ for weir flow in fitting the discharge capacity curves. In this study, $1/m$ is set to range from 0.5 to 2 due to the use of multiple discharge structures.

Equations (7) and (8) are combined, and let $K = K_1/K_2^n$:

$$V = KQ_{out}^n \tag{9}$$

Q_{in} is the inflow hydrograph, which is assumed to have a rectangular shape (Volpi & Flori 2014):

$$Q_{in} = \begin{cases} Q_{max} & t \le t_i \\ 0 & t > t_i \end{cases} \tag{10}$$

where Q_{in} is the inflow hydrograph, and Q_{max} is the peak flow.

The water balance equation is:

$$Q_{in} - Q_{out} = \frac{dV}{dt} \tag{11}$$

Then, the following nonlinear differential equation can be obtained:

$$\frac{dQ_{out}}{dt} = \frac{1}{Kn} Q_{in} Q_{out}^{1-n} - \frac{1}{Kn} Q_{out}^{2-n} \tag{12}$$

Equation (12) can be solved using polynomial integration, approximate polynomial integration, and incomplete beta function. Here, the approximate polynomial integration method (Bender & Orszag 1999) is used to derive the relationships among flood peak, flood volume, design discharge, and maximum storage capacity of the reservoir due to its relatively simple expression.

$$\frac{dQ_{out}}{dt} + \frac{1}{Kn} Q_{out}^{2-n} = \frac{1}{Kn} Q_{in} Q_{out}^{1-n} \tag{13}$$

$$\frac{Q_{max}}{Kn} dt = Q_{out}^{n-1} \left(\frac{1}{1 - \dfrac{Q_{out}}{Q_{max}}} \right) dQ_{out} \tag{14}$$

Series expansion:

$$\frac{Q_{max}}{Kn} dt = Q_{out}^{n-1} \left(\left(\frac{Q_{out}}{Q_{max}}\right)^0 + \left(\frac{Q_{out}}{Q_{max}}\right)^1 + \left(\frac{Q_{out}}{Q_{max}}\right)^2 + \cdots + \left(\frac{Q_{out}}{Q_{max}}\right)^r \right) dQ_{out} \quad (r \to \infty) \tag{15}$$

$$\frac{Q_{max}}{Kn} dt = \left(Q_{out}^{n-1} + \frac{Q_{out}^{n}}{Q_{max}} + \frac{Q_{out}^{n+1}}{Q_{max}^2} + \cdots + \frac{Q_{out}^{n-1+r}}{Q_{max}^{t}} \right) dQ_{out} \quad (r \to \infty) \tag{16}$$

Integration of both sides yields:

$$\frac{Q_{max} t}{Kn} = \frac{Q_{out}^{n}}{n} + \frac{Q_{out}^{n+1}}{(n+1)Q_{max}} + \frac{Q_{out}^{n+2}}{(n+2)Q_{max}^2} \cdots + \frac{Q_{out}^{n+r}}{(n+r)Q_{max}^r} \quad (r \to \infty) \tag{17}$$

$$\frac{Q_{max} t}{Kn} = \frac{Q_{out}^{n}}{n} + Q_{out}^{n} \sum_{i=1}^{\infty} \frac{Q_{out}^{i}}{(n+i)Q_{max}^{i}} \tag{18}$$

$1/n + i$ can be approximated as $1/i$:

$$\frac{Q_{max} t}{Kn} = \frac{Q_{out}^{n}}{n} + Q_{out}^{n} \sum_{i=1}^{\infty} \frac{1}{i} \left(\frac{Q_{out}}{Q_{max}}\right)^i \tag{19}$$

$$Q_{max} t = K Q_{out}^{n} \left(1 - n \ln\left(1 - \frac{Q_{out}}{Q_{max}}\right)\right) \tag{20}$$

When $t = t_i$, Q_{out} in Equation (20) gets the maximum value $Q_{out,max}$:

$$Q_{max}t_i = KQ^n_{out,max}\left(1 - n\ln\left(1 - \frac{Q_{out,max}}{Q_{max}}\right)\right) \tag{21}$$

The flood volume $W = Q_{max}t_i$ can be obtained from Equation (10), and the maximum reservoir volume $V_{max} = KQ^n_{out,max}$ during the routing process can be obtained from Equation (9):

$$W = V_{max}\left(1 - n\ln\left(1 - \frac{Q_{out,max}}{Q_{max}}\right)\right) \tag{22}$$

where W is the flood volume, and V_{max} is the maximum reservoir volume during the routing process.

The maximum reservoir discharge $Q_{out,max}$ can be obtained for each inflow hydrograph using Equation (22). For a given return period T of the maximum discharge, the maximum design discharge Q_D can be described as:

$$Q_D = F^{-1}_{Q_{out,max}}\left(1 - \frac{1}{T}\right) \tag{23}$$

where $\mu = 1$.

According to the mathematical model described in Section 2, Q_D is the design parameter, and Q_{in} and W are random variables which take into consideration the hydrological loads acting on the structure. The structure function is shown in Equation (22). Then, the probability distribution function of Q_D can be obtained, and the boundary of the integration region D_z is:

$$W = \gamma(Q_D) = V_{max}\left(1 - n\ln\left(1 - \frac{Q_{out,max}}{Q_{max}}\right)\right) \tag{24}$$

Let the maximum design discharge be the same as the maximum discharge in Equation (24), and thus $Q_{out,max} = Q_D$. The function has two asymptotes: $W \to \infty$ as $Q_{in} \to Q_{out,max}$ and $Q_{in} \to \infty$ as $W \to V_{max}$. It follows from Equation (22) that the same $Q_{out,max}$ can be derived from the limiting cases of the inflow hydrograph, which is characterized by an infinite peak flow but with a constant flood volume equal to V_{max}, or by an infinite flood volume but with a constant peak flow equal to $Q_{out,max}$.

Equation (24) can be rewritten as:

$$F_{Q_{out,max}}(q_{max}) = \int_{Q_{out,max}}^{\infty}\int_{V_{max}}^{\gamma(Q_{in})} f_{Q_{in},W}(q_i, w_i)dWdQ_{in} + F_W(V_{max}) + F_{Q_{in}}(Q_{out,max}) - \int_{-\infty}^{Q_{out,max}}\int_{-\infty}^{V_{max}} f_{Q_{in},W}(q_i, w_i)dWdQ_{in} \tag{25}$$

$F_{Q_{out,max}}(q_{max})$ is the probability distribution function of the maximum design discharge $Q_{out,max}$; $f_{Q_{in},W}(q_{in}, w_i)$ is the joint probability density function of random variables Q_{in} and W, where $F_{Q_{in}}$ and F_W are their marginal distribution functions, respectively. When the design variable $Q_{out,max}$ is independent of one of the two random variables (e.g., W), the boundary of D_Z can be described as a straight line $q_i = Q_{out,max}$, and thus $F_{Q_{out,max}} = F_{Q_{in}}$.

It is important to note that the reservoir discharge-water level relationship remains approximately constant during the routing process, and both the maximum water level at the dam and the maximum reservoir volume can be achieved once the maximum discharge is achieved, which strictly corresponds to the structure return period.

4. CASE STUDY

The Xin'an reservoir is located on the tributary of the Qiantang River in Tongguan Xia, Zhejiang Province, China Figure 2, and the upstream river is 323 km long with a drainage area of 10,442 km^2. This reservoir is intended for flood control and power generation, and it has a multiyear regulation capacity with a normal pool level of 108 m and a storage capacity of 17.84 billion m^3. The main discharge structure of the reservoir is the spillway with nine sluice gates, where the net width of the spillway is 117 m, and the elevation of the spillway crest is 99 m. The flood data of the Luotongbu Station, the inflow observation station of the reservoir, for the period 1961–2008 were collected, from which the annual maximum flood peak flow Q_{max} and the annual maximum 10-d flood volume W_{10} were extracted.

4.1. Marginal and joint distributions

The marginal distribution functions of Q_{max} and W_{10} are derived. The generalized extreme value (GEV) distribution is used in this study as the marginal distribution function due to its wide use and good performance in flood frequency analysis. The Gumbel copula is considered to best represent the relationship between peak and volume (Zhang & Singh 2006). Three copulas (G–H, Frank, and Clayton) from the Archimedean copula family are used in this study to simulate Q_{max}–W_{10} joint distribution, and the maximum likelihood method is used to estimate the parameter θ. The optimal copula function is selected using Akaike information criterion (AIC), Bayesian information criterion (BIC) and root-mean-square error (RMSE), and the smaller the values of RMSE, AIC and BIC, the better the fitting of the copula function. Table 1 shows that the smallest AIC and BIC values are obtained by G–H copula, but its RMSE value is slightly larger than that of Frank copula. Thus, the G–H copula is used to describe the Q_{max}–W_{10} joint distribution in this study.

4.2. Structure return period analysis

In this study, the maximum water level in the reservoir above the spillway crest H_{max} is taken as the structure design variable. A nonlinear reservoir model is established, and then the structure return period is calculated and compared with that obtained based on a linear reservoir model.

4.2.1. Structure return period based on the nonlinear reservoir model

(1) Reservoir volume-water level relationship

The elevation of the spillway crest of the Xin'an reservoir (99 m) is taken as the reference, at which the water level at the dam is $H = 0$ m and the reservoir volume is $V = 0$ m^3. Then, the reservoir volume-water level relationship is fitted and described as follows:

$$V = 4.3341H^{1.1} \tag{26}$$

(2) Reservoir discharge-water level relationship

The water of the Xin'an reservoir is discharged mainly through an overflow spillway with a typical WES profile. This permits a free outflow without side contraction and marching of water head, and the water level at the dam–discharge relationship is fitted and described as follows:

$$Q_{out} = 264.31H^{3/2} \tag{27}$$

(3) Reservoir discharge–volume relationship

The discharge–reservoir volume relationship is obtained from Equations (26) and (27):

$$V = 0.07256Q_{out}^{11/15} \tag{28}$$

Table 1 | Parameters of the Q_{max}–W_{10} joint distribution

Functions	θ	AIC	BIC	RMSE
G–H copula	2.8506	−283.49	−278.16	0.0477
Clayton copula	2.6746	−279.73	−274.72	0.0499
Frank copula	10.5565	−254.15	−259.55	0.0386

(4) Structural function

Substitution of Equation (28) into the water balance Equation (11) yields:

$$W_{10} = 4.3341H_{\max}^{1.1}\left(1 - \frac{11}{15}\ln\left(1 - \frac{264.31H_{\max}^{3/2}}{Q_{\text{in}}}\right)\right) \tag{29}$$

which is also the expression of the boundary of the integration region D_Z of $H_{\max}$.

The probability of D_Z at a given $h_{\max}$ can be calculated from Equation (29), and then the structure return period is determined to be:

$$T_{n,s} = \frac{1}{1 - F_{H_{\max}}(h_{\max})} \tag{30}$$

4.2.2. Structure return period based on the linear reservoir model

The structure return period based on the linear reservoir model in which both reservoir volume-water level and discharge-water level relationships are assumed to be linear is also calculated following the same procedure and compared with that based on the nonlinear model (see Volpi & Flori (2014) and Balistrocchi *et al.* (2017) for details).

The linear reservoir water level–volume relationship can be simulated using the measured reservoir volume and water level data:

$$V = 5.5791H \tag{31}$$

The linear reservoir water level–discharge relationship is:

$$Q_{\text{out}} = 974.11H \tag{32}$$

The linear reservoir volume–discharge relationship is:

$$V = 0.00573Q_{\text{out}} \tag{33}$$

The structure function is:

$$W_{10} = -0.00573Q_{\text{in}}\ln\left(1 - \frac{974.11H_{\max}}{Q_{\text{in}}}()\right) \tag{34}$$

Then,

$$T_{l,s} = \frac{1}{1 - F_{H_{\max}}(h_{\max})} \tag{35}$$

4.3. Comparison of structure return periods between linear and nonlinear models

In this section, the structure return periods based on linear and nonlinear reservoir models are calculated and compared, and possible explanations of the difference are discussed.

Figure 3(a) shows the comparison of the maximum water level–return period relationships between linear and nonlinear models. It shows that when $T < 360$a, the $H_{\max}$ of the nonlinear model is larger than that of the linear model at the same return period; while the opposite trend is observed when $T > 360$a. As the $H_{\max}$ value increases, the T value of the nonlinear model increases more rapidly, which can be attributed to the more rapid change of $Q_{\text{out,max}}$ and $V_{\max}$ with $H_{\max}$, as shown in Figure 3(b).

Six representative return periods ($T = 10$, 50, 100, 200, 500, and 1,000a) are selected for further analysis, and Table 2 shows the $H_{\max}$, $Q_{\text{out,max}}$, and $V_{\max}$ at different return periods for linear and nonlinear models. It is found that the $H_{\max}$ values of the linear model are smaller than that of the nonlinear model at $T = 10$, 50, and 100a, while the opposite is true for the values of $Q_{\text{out,max}}$ and $V_{\max}$. The values of $H_{\max}$, $Q_{\text{out,max}}$, and $V_{\max}$ increase with the increase of the return period, which, however, is

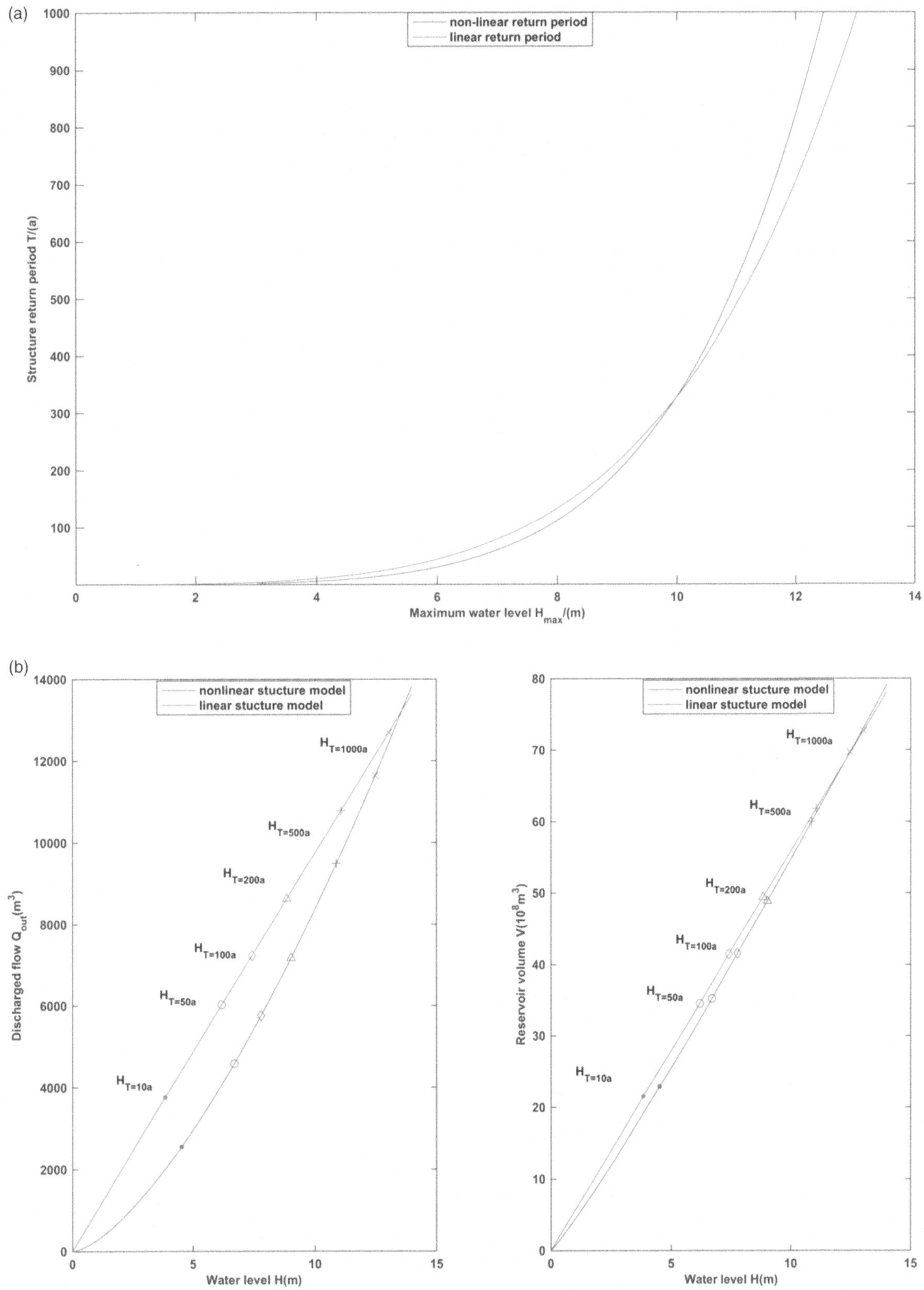

Figure 3 | Comparison of the relationships of the maximum water level with (a) the return period and (b) the maximum discharge and volume between linear and nonlinear models.

Table 2 | The maximum water level, maximum discharge, and maximum volume of the reservoir at different return periods for linear and nonlinear models

	Maximum water level (H_{max}/m)		Maximum discharge ($Q_{out,\ max}/(m^3/s)$)		Maximum reservoir volume ($V_{max}/10^8\ m^3$)	
Return period (*T*s/a)	Nonlinear model	Linear model	Nonlinear model	Linear model	Nonlinear model	Linear model
10	4.53	3.86	2,551.91	3,763.13	22.86	21.55
50	6.70	6.18	4,585.34	6,022.19	35.13	34.49
100	7.80	7.43	5,761.66	7,233.86	41.54	41.43
200	9.03	8.85	7,175.06	8,623.39	48.79	49.39
500	10.88	11.07	9,486.47	10,778.60	59.87	61.73
1,000	12.47	13.03	11,643.84	12,688.00	69.58	72.67

Table 3 | The maximum water levels at the dam at different return periods under non-stationary and stationary conditions based on the nonlinear storage/discharge model

	Non-stationary conditions					
Return period (*T*s/a)	1961	1971	1981	1991	2001	Stationary conditions
10	4.02	4.22	4.43	4.66	4.89	4.53
50	4.94	5.19	5.45	5.72	6.01	6.70
100	5.26	5.52	5.81	6.10	6.41	7.80
200	5.55	5.83	6.12	6.43	6.75	9.03
500	5.88	6.18	6.48	6.81	7.16	10.88
1,000	6.10	6.41	6.73	7.07	7.43	12.47

more pronounced for the linear model. This is because the discharge changes more rapidly with the water level for the nonlinear model, as shown in the $Q_{out,max}$–H curve. At a return period of 500 years, the values of H_{max}, $Q_{out,max}$, and V_{max} of the linear model are larger than those of the nonlinear model, indicating that flood risk is over-estimated by the linear model.

Figure 4 shows the comparison of the structure return period isolines at $T = 10$, 50, 100, 200, 500, and 1,000a between linear and nonlinear reservoir models, and gray dots are the 1,000 random combinations of flood peak and volume generated using the G–H copula. It is important to note that any flood event on the same isoline would produce the same maximum reservoir volume, maximum water level at the dam, and maximum discharge, and thus have the same impact on the reservoir. However, given their different assumptions about the reservoir volume-water level and discharge-water level relationships, the boundary of the supercritical region differs substantially between linear and nonlinear reservoir models, resulting in substantial differences in the isolines at the same return period. It is seen that at a return period of $T = 10$, 50, 100, and 200a, the discharge for the linear reservoir model is higher compared with that for the nonlinear reservoir model, and the nonlinear reservoir model is more sensitive to floods with high flood peak but low flood volume; whereas at a return period of $T = 500$a and $T = 1{,}000$a at which the maximum water level at the dam is high, the discharge and volume are approximately the same between linear and nonlinear reservoir models, and as a result the isolines are approximately coincident with each other.

It is thus concluded that reservoir discharge and volume would be increased to different extents with increasing water level for linear and nonlinear reservoir models due to their differences in the reservoir volume-water level and discharge-water level curves, which consequently have an effect on the structure return period. In addition, the reservoir volume-water level and discharge-water level curves can also have an effect on the boundary of the supercritical region, thus resulting in differences in the isolines at the same return period between linear and nonlinear reservoir models.

4.4. Comparison of different return periods

For the Kendall return period, the supercritical region is defined as the region enveloped by the isoline of the joint return period. Here, the structure return period which is applicable to the flood control structures and the Kendall return period

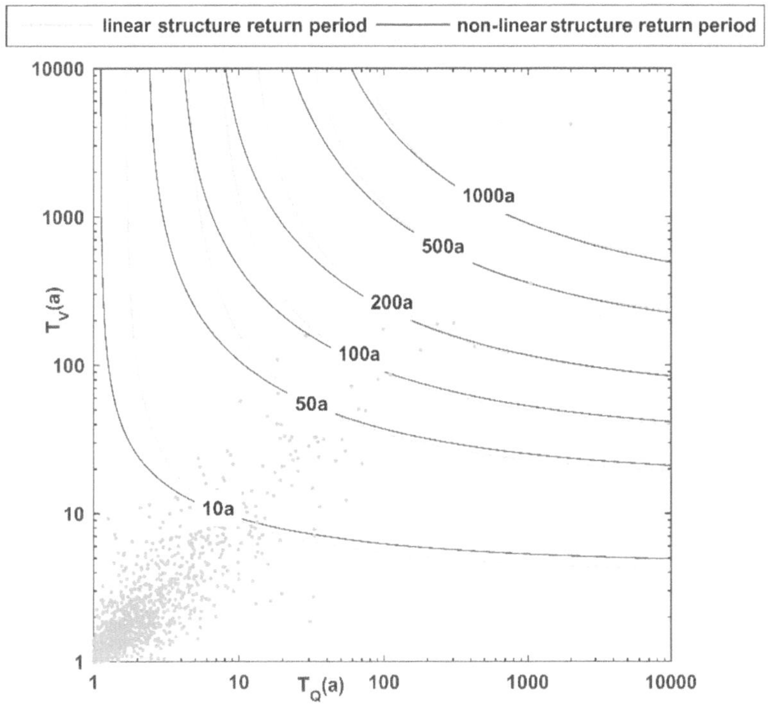

Figure 4 | The isolines of the structure return periods for linear and nonlinear models.

which is applicable to natural occurrence of floods are compared. The isolines of the structure return period and the Kendall return period at a return period of $T = 10$, 50, 100, 200, 500, and 1,000a are shown in Figure 5, and gray dots in Figure 5 are the 1,000 random combination of flood peak and volume generated using the G–H copula.

It shows that the Kendall return period isoline is symmetric about the straight line with an angle of 45°, as it is based on the joint return period; while the symmetry axis of the structure return period isoline is more inclined to the y-axis, which is affected by the reservoir structure such that a flood event with a larger peak flow can pose a greater risk to the reservoir structure.

There are an infinite number of peak–volume pairs at the same Kendall return period, which can produce different reservoir responses and thus pose different levels of risk to the flood control structures. The Kendall return period of $T = 10$a is analyzed. It is noted that the isoline with a Kendall return period of $T = 10$a intersects that with a structure return period of 10, 50, and 100a, which are denoted as points A, B, C, and D. The occurrence probability of floods at points A, B, C, and D is 10a; the return period of the maximum water level at the dam for peak–volume pairs at points A and D is also 10a; while that at points B and C is 50 and 100a, respectively. Thus, the maximum water level in response to peak–volume pairs on the left side of point A and the right side of point D on the Kendall return period isoline of $T = 10$a is higher than 10a, while that between points A and D is lower than 10a. It is concluded that the reservoir would respond differently to different peak–volume pairs at the same Kendall return period, and thus flood risk is likely to be substantially underestimated in some circumstances. At a Kendall return period of $T = 10$a, the maximum reservoir response is 196.34a and the minimum reservoir response is 8.96a.

The Kendall return period is often used in bivariate design, and the maximum likelihood method, the most likely composition method, and the same frequency method are used for flood peak and volume estimation. The bivariate design based on the Kendall return period accounts for the natural occurrence of floods without considering the effects of hydraulic structures during the routing process, which can lead to over- or underestimation of flood risk from the perspective of hydraulic structure safety.

5. DISCUSSION

Although the nonlinear reservoir model has been successfully applied to the Xin'an Reservoir, other reservoirs may show quite different reservoir volume-water level and discharge-water level relationships ($V = K_1 H^{n/m}$ and $Q_{\text{out}} = K_2 H^{1/m}$). Thus, the effects of the four parameters (K_1, n/m, K_2, and $1/m$) on the results are analyzed to verify the applicability of the proposed model to other hydraulic structures.

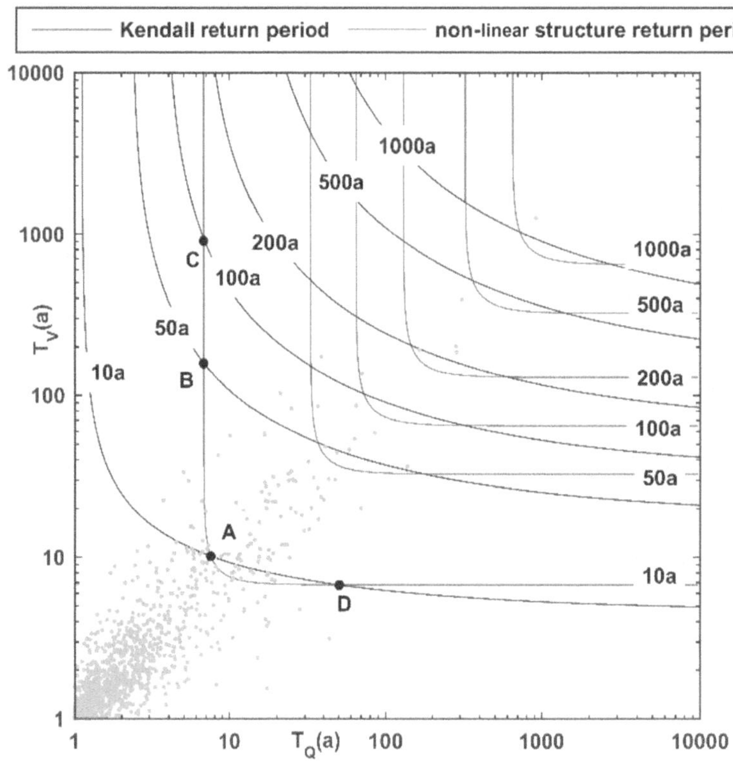

Figure 5 | Comparison of the nonlinear structure return period and the Kendall return period.

5.1. Effects of parameters of the discharge capacity curves

The discharge capacity of a reservoir depends largely on the spillway length and discharge regime, which correspond to K_2 and $1/m$ in Equation (8), respectively. Specifically, $1/m$ indicates the discharge capacity of the reservoir under different discharge regimes, which can be affected not only by the structure, such as sluice gate and overflow weir, but also by flow patterns such as orifice and weir outflow through the sluice gate. In general, the larger the discharge capacity is, the larger the $1/m$ value will be. K_2 indicates the spillway length, and the larger the spillway length is, the larger the K_2 value will be. Here, a constant reservoir volume-water level relationship is considered ($V = H^{1.1}$), and the effects of K_2 (100, 200, and 300) and $1/m$ (0.5, 1, 1.5, and 2) on the discharge capacity ($Q_{out} = K_2H^{1/m}$) are investigated.

Figure 6 shows the relationships between the structure return period (T) and the maximum water level ((H_{max})) for different discharge curves. It shows that at the same structure return period T, an increase in K_2 or $1/m$ can lead to an increase in the discharge capacity and consequently a decrease in H_{max}. However, it is important to note that there is a nonlinear relationship between T and H_{max}, and T increases at an increasing rate with the increase of H_{max}.

As $1/m$ indicates the discharge capacity of the reservoir under different discharge regimes, it can have a significant effect on the return period. For instance, as $1/m$ is increased by 100, 50, and 33.3% from $1/m_1 = 0.5$ ($Q = 100H^{0.5}$) to $1/m_2 = 1(Q = 100H)$, $1/m_3 = 1.5$ ($Q = 100H^{1.5}$), and $1/m_4 = 2$ ($Q = 100H^2$) at $K_2 = 100$, the H_{max} is decreased by 10.9, 23.4, and 34.3% from $h_1 = 19.49$ m to $h_2 = 17.38$ m, $h_3 = 13.30$ m, and $h_4 = 8.74$ m at $T = 10$a, respectively. Under the same K_2 conditions, the structure return period for the same H_{max} would increase at an increasing rate with the increase of $1/m$. This is because the discharge capacity of the reservoir is increased as the H_{max} increases, and thus the larger the $1/m$ value is, the higher the increase rate of the discharge capacity of the reservoir will be.

It is also noted that K_2, which is used to indicate the increase or decrease in the spillway length, can also have an effect on the $H_{max} \sim T$ relationship. For instance, as K_2 is increased by 100 and 50% from $k_{2,1} = 100$ ($Q = 100H^{0.5}$) to $k_{2,2} = 200$ ($Q = 200H^{0.5}$), and $k_{2,3} = 300$ ($Q = 300\mathrm{H}^{0.5}$) at $T = 10$a and $1/m = 0.5$, the H_{max} is decreased by 9.1 and 8.0% from $h_1 = 19.49$ m to $h_4 = 17.72$ m and $h_5 = 16.30$ m, respectively. Under the same $1/m$ conditions, the structure return period for the same H_{max} would also increase at an increasing rate with the increase of K_2. Again, this is because the discharge capacity of the reservoir is increased as the H_{max} increases, and thus the larger the K_2 value is, the higher the increase rate of the discharge capacity of the reservoir will be.

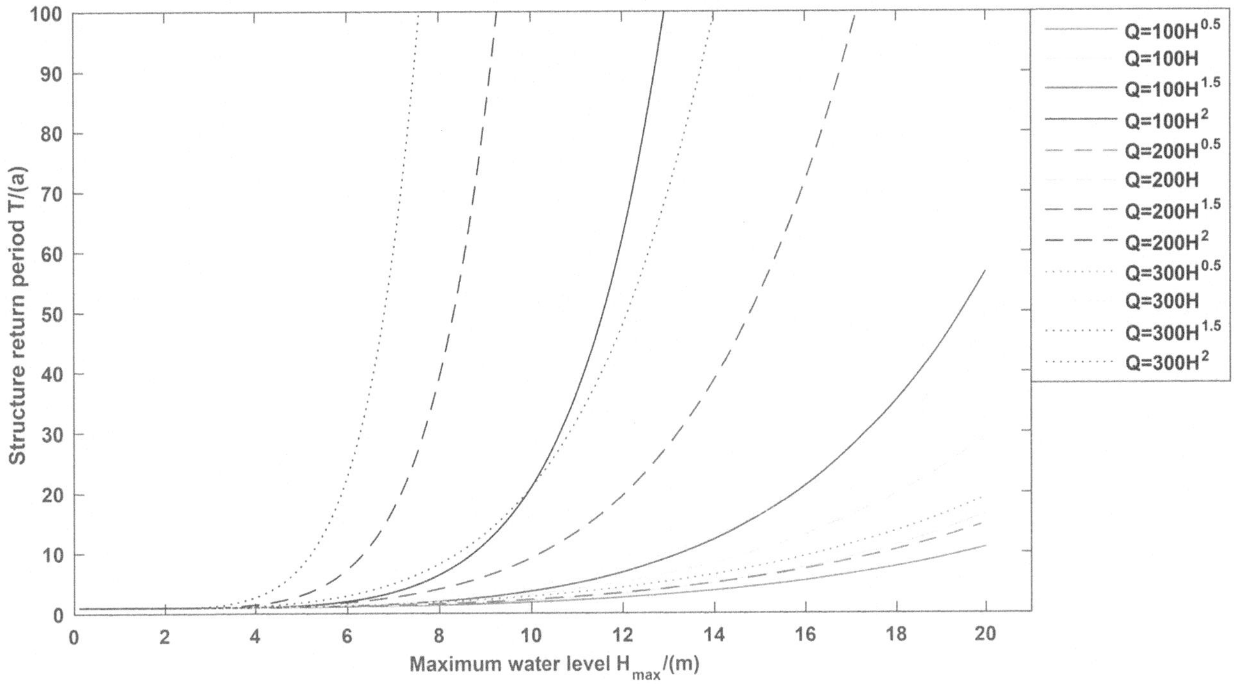

Figure 6 | The relationship between the structure return period and the maximum water level for different parameters of the discharge capacity curves.

5.2. Effects of parameters of reservoir volume and water level curves

The relationship between reservoir water level and its volume is closely associated with the geographical location and characteristics of the reservoir, which can also have an effect on the structure return period. Under the conditions of $Q_{out} = 250H^{1.5}$, the effects of parameters K_1 ($K_1 = 1$, 1.5, 2, or 3) and n/m ($n/m = 1$, 1.2, 1.5, or 2) in $V = K_1H^{n/m}$ are discussed.

Figure 7 shows the $H_{max} \sim T$ relationships for different reservoir volumes. At the same structure return period, an increase in K_1 or n/m can lead to an increase in the storage capacity and consequently a decrease in H_{max}. It is also note that the relationship between T and H_{max} is nonlinear, and T increases at an increasing rate with the increase of H_{max}.

Note that as the storage capacity is mainly affected by the geographical location of the reservoir, it would be much more difficult to change than reservoir discharge capacity. Nevertheless, changes in reservoir storage capacity can have a substantial impact on the safety of hydraulic structures. For instance, under the conditions of $n/m = 1.1$ ($V = H^{1.1}$, $V = 1.5H^{1.1}$, $V = 2H^{1.1}$ and $V = 3H^{1.1}$), an increase of K_1 by 50, 100, and 200% results in a decrease in H_{max} by 12.1, 22.4, and 37.4% at $T = 10$a, respectively. Under the same n/m conditions, the structure return period for the same H_{max} would increase at an increasing rate with the increase of K_1. Again, this is because the storage capacity of the reservoir is increased as the H_{max} increases, and thus the larger the K_1 value is, the higher the increase rate of the storage capacity of the reservoir will be.

For curves with $K_1 = 1$ ($V = H$, $V = H1.1$, $V = H1.2$, $V = H1.5$, and $V = H2$), an increase of n/m by 10, 20, 50, and 100% results in a decrease in H_{max} by 5.6, 12.6, 33.0, and 54.2% at $T = 10$a, respectively. Under the same K_1 conditions, the structure return period for the same H_{max} would increase at an increasing rate with the increase of n/m. Again, this is because the storage capacity of the reservoir is increased as the H_{max} increases, and thus the larger the n/m value is, the higher the increase rate of the storage capacity of the reservoir will be.

5.3 Analysis of the Xin'an reservoir under stationary and non-stationary conditions

Figure 8 shows the time series of the annual maximum flood peak flow Q_{max} and the annual maximum 10-d flood volume W_{10} recorded at the Luotongbu Station for the 48-year period 1961–2008. It is seen that both Q_{max} and W_{10} time series show an increasing trend for the period 1961–2008.

The annual flood peak and the annual maximum 10-d flood volume records at the Luotongbu Station for the period 1961–2008 were fitted based on the GAMLSS model and the parameters were estimated. Figure 9 shows the evolution of Q_{max} and W_{10} at a return period of 1,000, 500, 200, 100, 50, and 10 years for the period 1961–2008.

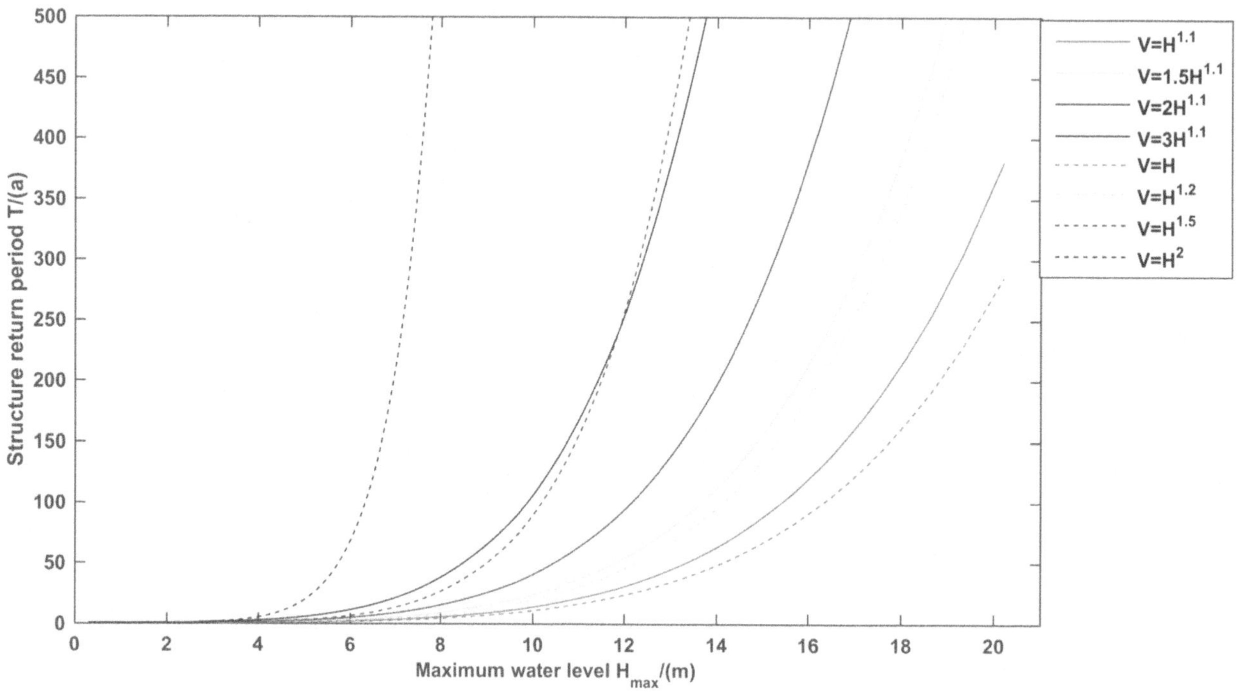

Figure 7 | The relationship between the structure return period and the maximum water level for different parameters of reservoir volume and water level curves.

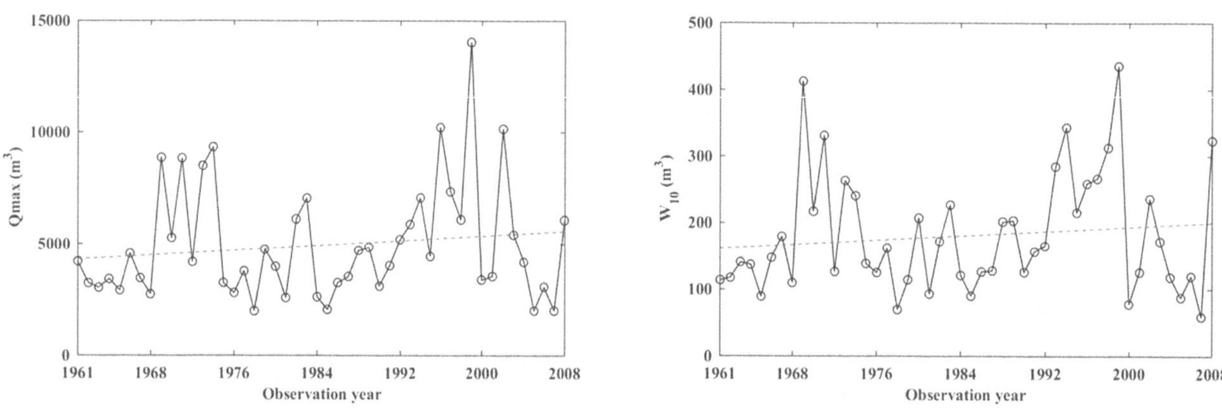

Figure 8 | Time series of annual maximum flood peak flow Q_{max} and annual maximum 10-d flood volume W_{10} for the period 1961–2008.

The distribution functions of the peak flow and volume in 1961, 1971, 1981, 1991, and 2001 under non-stationary conditions are selected, and the non-linear model is applied to the Xin'an reservoir to verify its applicability under non-stationary conditions.

Figure 10 shows the relationships between the maximum water level at the dam and the return period under non-stationary and stationary conditions, and Figure 11 shows the marginal distributions of annual maximum flood peak flow Q_{max} and the annual maximum 10-d flood volume W_{10} under non-stationary and stationary conditions. It is seen from Figure 10 that under non-stationary conditions, the maximum water level at the dam H_{max} for the same return period also shows an increasing trend for the period 1961–2008; while the return period T for the same year increases at a higher rate with the increase of the maximum water level at the dam. However, it should also be noted that the increasing rate is higher under non-stationary conditions than that under stationary conditions as a result of the marginal distributions of annual maximum flood peak flow and annual maximum 10-d flood volume. As shown in Figure 11, higher flood peaks are more likely to occur under non-stationary conditions, which can cause more severe damage to the reservoir structure and consequently the return period.

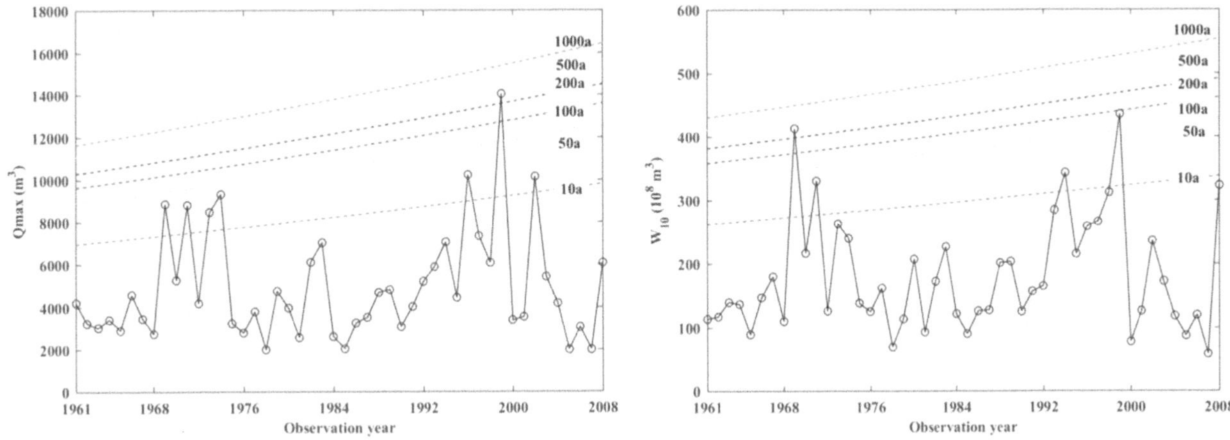

Figure 9 | The evolution of annual maximum flood peak flow Q_{max} and annual maximum 10-d flood volume W_{10} at a return period of 1,000, 500, 200, 100, 50, and 10 years for the period 1961–2008.

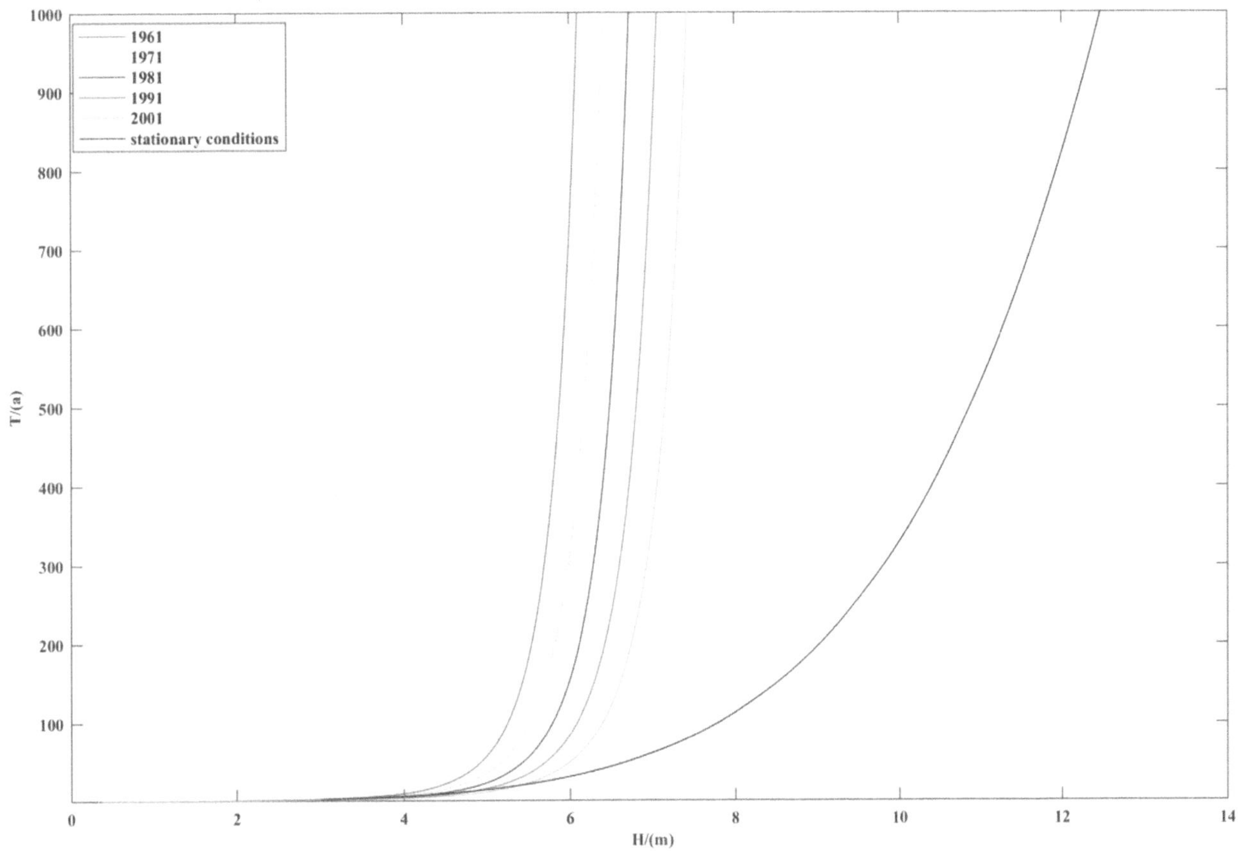

Figure 10 | The relationships between the maximum water level at the dam and the return period under non-stationary and stationary conditions.

Six representative return periods ($T = 10$, 50, 100, 200, 500, and 1,000a) are considered for comparison Table 3. Different Q_{max}–W_{10} joint distributions are used to obtain the nonlinear return periods under non-stationary conditions, and the increasing trend of Q_{max} and W_{10} for the period 1961–2008 implies that there is a higher probability of the joint occurrence of large Q_{max} and W_{10}. Under conditions of constant reservoir volume-water level and discharge-water level relationships, a higher

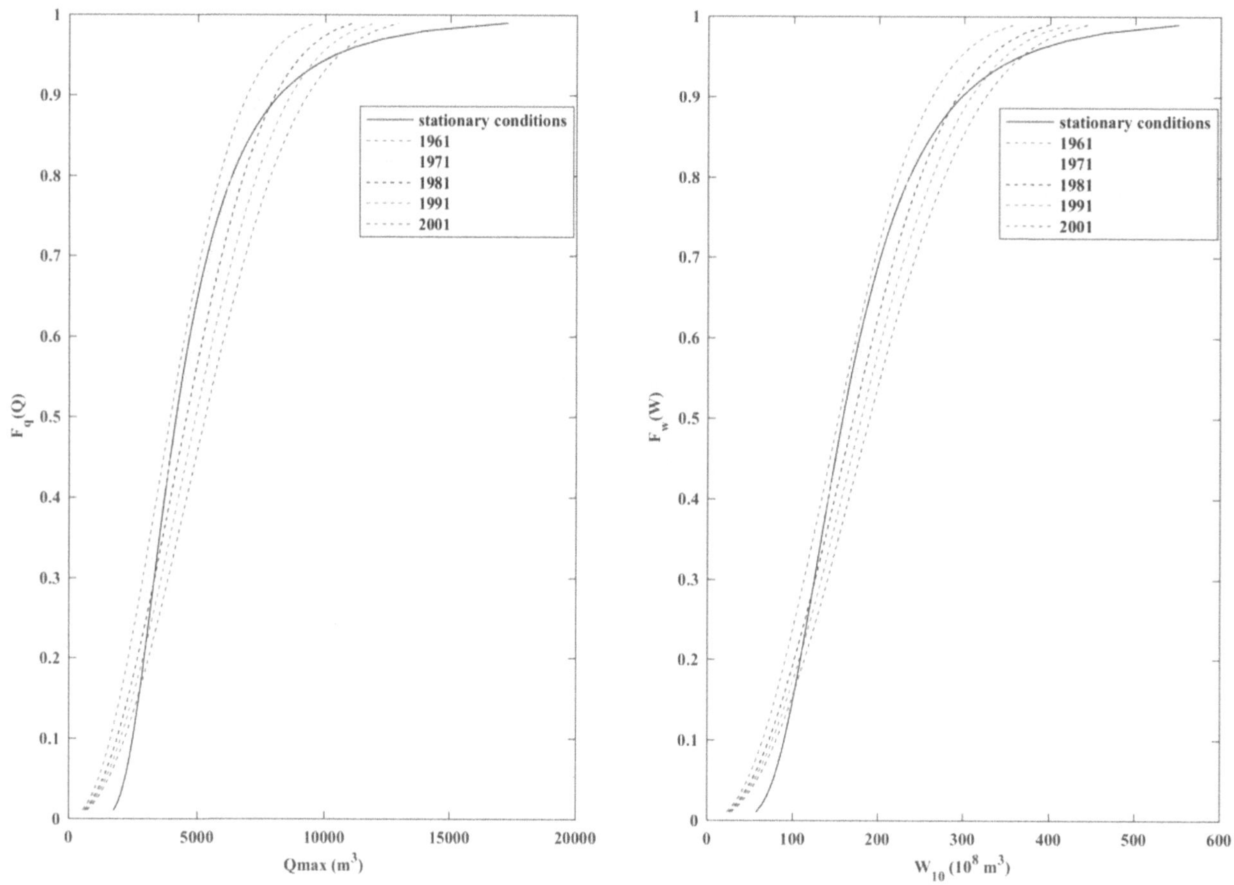

Figure 11 | The marginal distributions of annual maximum flood peak flow Q_{max} and the annual maximum 10-d flood volume W_{10} under non-stationary and stationary conditions.

maximum water level at the dam H_{max} is more likely to occur, which in turn may cause more severe damage to the reservoir. At a return period of $T = 10a$, the probability of the Q_{max}–W_{10} joint distribution under non-stationary conditions is similar to that under stationary conditions. Thus, under conditions of constant reservoir volume-water level and discharge-water level relationships, there would be no substantial difference in the return period between non-stationary and stationary conditions. However, the probability of the joint occurrence of large Q_{max} and W_{10} would increase as the return period increases under non-stationary conditions because of the effects of the marginal distributions of annual maximum flood peak flow and annual maximum 10-d flood volume. In this case, the return periods obtained under stationary conditions are more conservative.

6. CONCLUSIONS

In this study, a general nonlinear reservoir model was proposed for the design and risk assessment of hydraulic structures in a bivariate context. A case study was then performed with the Xin'an reservoir, and the nonlinear structure return period was calculated and compared with the linear structure return period and the Kendall return period. The main conclusions are summarized as follows.

1. The supercritical region D_Z plays a determinant role in the design and risk assessment of hydraulic structures, and its boundary is affected by the structure and hydrological load acting on the structure. Thus, it is imperative to consider the structure and the way it interacts with the hydrological load in determining the structure function. In this study, a non-linear reservoir model is established, and the relationship between structure and hydrological load is determined using the approximate polynomial integration method and then the supercritical region D_Z is obtained. Unlike the linear reservoir model, nonlinear reservoir volume-water level and discharge-water level relationships are assumed in this study, thus making it possible to capture the routing processes of the reservoir more accurately and flexibly.

2. The maximum water level at the dam is taken as the structure design variable, and the structure return period based on the nonlinear reservoir model is calculated and compared with that based on the linear reservoir model. Given their different assumptions about the reservoir volume-water level and discharge-water level relationships, the boundary of the supercritical region differs substantially between linear and nonlinear reservoir models, resulting in substantial differences in the structure return period. The change rates of discharge and reservoir volume with reservoir water level may also differ between linear and nonlinear reservoir models, and thus they show different levels of sensitivity to floods with different peak–volume combinations and thus lead to differences in the structure return period.
3. The structure-based multivariate return period is distinctively different from the hydrological event-based multivariate return period. The structure return period is more applicable than the Kendall return period to flood control structures as it takes into consideration the interactions of the structure and hydrological load acting on the structure. However, it is important to note that the Kendall return period can lead to over- or underestimation for hydraulic structures in multivariate flood frequency analysis.
4. For the nonlinear reservoir model, the effects of K_1, n/m, K_2, and $1/m$ involved in the reservoir volume-water level and discharge-water level relationships ($V = K_1H^{n/m}$ and $Q_{out} = K_2H^{1/m}$) are investigated to verify the applicability of the proposed model. It is found that an increase in K_1 or n/m leads to an increase in reservoir storage capacity; while an increase in K_2 or $1/m$ leads to an increase in reservoir discharge capacity. Different reservoirs would respond differently to the same flood event due to their different storage and discharge capacity, which can have an effect on the structure return period.

However, as it is difficult to derive the general solution of the nonlinear differential equation, the inflow hydrograph is simplified to be a constant process in this study, which may make the results less accurate. Efforts should also be made to better describe the fluctuation of the inflow hydrograph and to investigate the effects of the operation rules on the routing processes of the reservoir.

ACKNOWLEDGEMENTS

This research is funded by the National Natural Science Foundation of China (52079040), the Fundamental Research Funds for the Central Universities (B210201023), and Jiangsu Water Science and Technology Project (2019027 and 2020065).

DATA AVAILABILITY STATEMENT

All relevant data are included in the paper or its Supplementary Information.

REFERENCES

Balistrocchi, M., Orlandini, S., Ranzi, R. & Bacchil, B. 2017 Copula-based modeling of flood control reservoirs. *Water Resour. Res.* **53**, 9883–9900. https://doi.org/10.1002/2017WR021345.

Bender, C. M. & Orszag, S. A. 1999 *Advanced Mathematical Methods for Scientists and Engineers*. Springer, New York, NY.

Chebana, F. & Ouarda, T. 2011 Multivariate quantiles in hydrological frequency analysis. *Environmetrics* **22**, 63–78. https://doi.org/10.1002/env.1027.

De Michele, C., Salvadori, G., Canossi, M., Petaccia, A. & Rosso, R. 2005 Bivariate statistical approach to check adequacy of dam spillway. *J. Hydrol. Eng.-ASCE* **10**, 50–57. https://doi.org/10.1061/(ASCE)1084-0699(2005)10:1(50).

Favre, A.-C., El Adlouni, S., Perreault, L., Thiemonge, N. & Bobee, B. 2004 Multivariate hydrological frequency analysis using copulas. *Water Resour. Res.* **40**, 1–12. https://doi.org/10.1029/2003WR002456.

Klein, B., Pahlow, M., Hundecha, Y. & Schumann, A. 2010 Probability analysis of hydrological loads for the design of flood control systems using copulas. *J. Hydrol. Eng.-ASCE* **15**, 360–369. https://doi.org/10.1061/(ASCE)HE.1943-5584.0000204.

Mediero, L., Jiménez-Álvarez, A. & Garrote, L. 2010 Design flood hydrographs from the relationship between flood peak and volume. *Hydrol. Earth Syst. Sci.* **14**, 2495–2505. https://doi.org/10.5194/hess-14-2495-2010.

Nelsen, R. B. 2006 *An Introduction to Copulas*, 2nd edn. Springer, New York, NY.

Requena, A., Mediero, L. & Garrote, L. 2013 A bivariate return period based on copulas for hydrologic dam design: accounting for reservoir routing in risk estimation. *Hydrol. Earth Syst. Sci.* **17**, 3023–3038. https://doi.org/10.5194/hess-17-3023-2013.

Salvadori, G. & De Michele, C. 2004 Frequency analysis via copulas: theoretical aspects and applications to hydrological events. *Water Resour. Res.* **40**, 1–17. https://doi.org/10.1029/2004WR003133.

Salvadori, G., De Michele, C. & Durante, F. 2011 On the return period and design in a multivariate framework. *Hydrol. Earth Syst. Sci.* **15**, 3293–3305. https://doi.org/10.5194/hess-15-3293-2011.

Salvadori, G., Durante, F. & De Michele, C. 2013 Multivariate return period calculation via survival functions. *Water Resour. Res.* **49**, 2308–2311. https://doi.org/10.1002/wrcr.20204.

Shiau, J., Wang, H. & Tsai, C. 2006 Bivariate frequency analysis of floods using copulas. *J. Am. Water Resour. Assoc.* **42**, 1549–1564. https://doi.org/10.1111/j.1752-1688.2006.tb06020.x.

Volpi, E. & Fiori, A. 2014 Hydraulic structures subject to bivariate hydrological loads: return period, design, and risk assessment. *Water Resour. Res.* **50**, 885–897. https://doi.org/10.1002/2013WR014214.

Yan, L., Xiong, L., Guo, S., Xu, C.-Y., Xia, J. & Du, T. 2017 Comparison of four nonstationary hydrologic design methods for changing environment. *J. Hydrol.* **551**, 132–150.

Yan, L., Xiong, L., Luan, Q., Jiang, C., Yu, K. & Xu, C.-Y. 2020 On the applicability of the expected waiting time method in nonstationary flood design. *Water Resour. Manag.* **34**, 2585–2601.

Zhang, L. & Singh, V. P. 2006 Bivariate flood frequency analysis using the copula method. *J. Hydrol. Eng.-ASCE* **11**, 150–164. https://doi.org/10.1061/(ASCE)1084-0699(2006)11:2(150).

First received 27 July 2021; accepted in revised form 31 July 2021. Available online 1 September 2021

doi: 10.2166/nh.2021.090

Hydro-environmental response to the inter-basin water resource development in the middle and lower Han River, China

Junhong Zhang[a,b,*], Liquan Guo[a], Tao Huang[c], Dongdong Zhang[d], Zhimin Deng[e], Linshuang Liu[f] and Tao Yang[g]

[a] College of Resources and Environmental Science, South-Central University for Nationalities, Wuhan 430074, China
[b] China Institute of Water Resources and Hydropower Research, Beijing 100038, China
[c] Guangxi Water & Power Design Institute Co., Ltd., Naning 530023, China
[d] Bureau of Hydrology, Changjiang Water Resources Commission, Wuhan 430010, China
[e] Changjiang Water Resources Protection Institute, Wuhan 430010, China
[f] Changjiang Waterway Institute of Planning, Design & Research, Wuhan, Hubei Province 430040, China
[g] Tianjin Research Institute for Water Transport Engineering, Tianjin 300456, China
*Corresponding author. E-mail: 2014059@mail.scuec.edu.cn

ABSTRACT

As one of the most important influencing factors, inter-basin water resource development has been exerting an increasingly evident impact on the hydro-environment of river basins. The Han River was selected as a case study to reveal the hydro-environmental response to China's inter-basin water resource development. The hydrological changes and water-quality variations resulted from the middle route of the South-to-North Water Transfer Project (SNWTP) and the Three Gorges Reservoir (TGR) operation were examined based on a hydro-environmental model. The results indicated that the runoff reduction is obvious after the SNWTP operation, and the low-flow duration significantly increased by 4–5 months. Consequently, the flow decrease significantly contributed to the water quality deterioration in the middle and lower Han River, while the Yangtze-Han Water Diversion Project (YHWDP) can not alleviate the situation completely. Moreover, the nutrient assimilative capacity decreased after water diversion, which agrees with the hydrological changes along the middle and lower Han River. The quantitative analysis performed in this study distinguishes the spatiotemporal variation in water quality variables using the integrated model. It provides insights into water quality management under the influences of inter-basin water resource development.

Key words: hydro-environmental changes assessment, hydro-environment model, inter-basin water resource development, the middle and lower Han River

HIGHLIGHTS

- A coupled hydro-environmental model was set up to evaluate the river health.
- Long low-flow duration was examined along the lower Han River.
- Water quality of the middle and lower Han River deteriorated after the TGR and SNWTP operation.
- SNWTP significantly contributed to the water environment deterioration.
- YHWDP can not completely offset the adverse environmental influences resulted from the SNWTP operation.

GRAPHICAL ABSTRACT

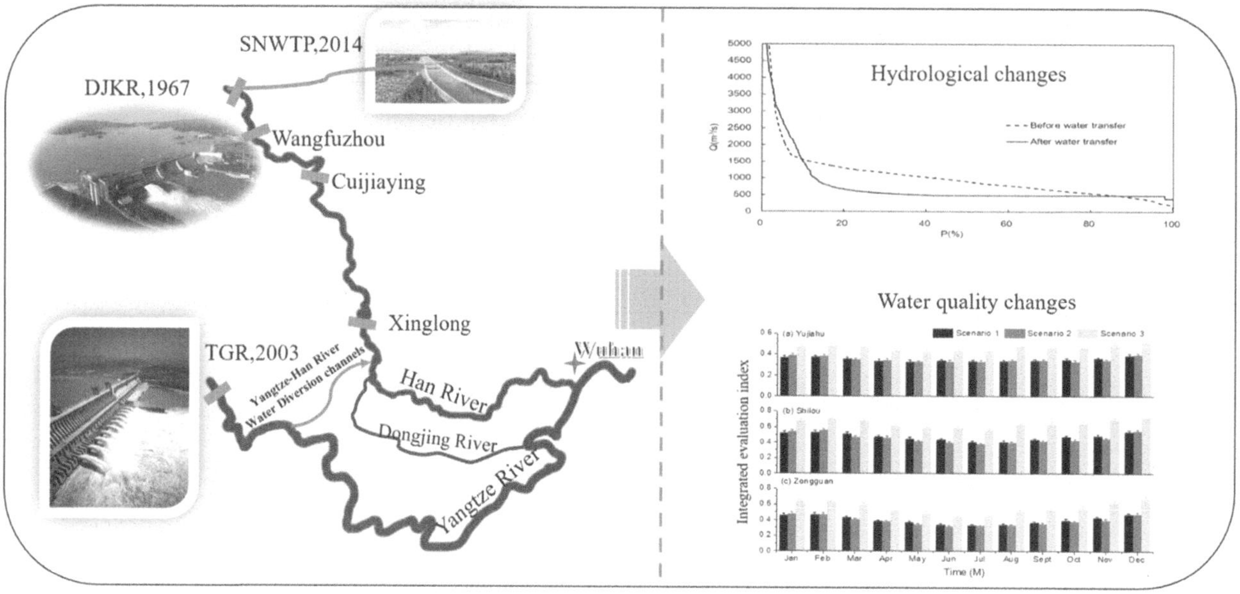

1. INTRODUCTION

The construction of water conservancy projects in river basins is an important way for human beings to utilize water resources comprehensively. However, the increasing intensity of human activities imposed on the natural river system seriously threatens the river water environment (Bond *et al.* 2008; Pruden *et al.* 2012; Koster *et al.* 2021). In the past decades, the hydro-environmental problems caused by strong inter-basin water resource development have attracted much attention (Palmer 2010; Vörösmarty *et al.* 2010; Bakker 2012; O'Connor *et al.* 2015). Construction and operation of large-scale water conservancy projects generally have continuously physical and ecological influences on the downstream river channels. Liechti *et al.* (2015) analyzed the impact of water conservancy construction on the hydrological situation of the Zambezi River in Africa and southern China based on historical data and found that the reservoirs have a significant impact on river flow fluctuations and seasonal hydrological changes. Bonacci & Oskorus (2010) conducted a study on the hydrological situation of the Drava River in Italy for nearly 30 years and clarified the effects of the construction and operation of the reservoirs on the water level, flow, and sediment transport in the lower reaches of the river. More than half of the large rivers in the world are under the influence of strong human activities, these rivers are facing higher pressures of hydro-environmental changes than the ones under natural conditions (Gurría 2009; Yang *et al.* 2017).

Sustainable development and integrated utilization of water resources while considering river health have become an increasingly important issue in the current research areas of water conservancy and hydro-environment (Scionti *et al.* 2018; Hao *et al.* 2021). At present, the construction of water conservancy projects in China has gradually shifted from traditional water conservancy to ecological water conservancy, more and more attention was paid to the coordinated development of engineering and environment, and more importance was attached to maintaining the ecological environment and safeguarding humanity (Cheng *et al.* 2019). To improve the nationwide utilization of water resources in China, the Three Gorges Reservoir (TGR) and the South-to-North Water Transfer Projects (SNWTP) were constructed and put into service successively. The TGR and SNWTP, up to now, are the largest hydraulic structure and the inter-basin water transfer project all over the world. Both projects have greatly improved the utilization of water resources and alleviated the water shortage in northern China, which contributed a great deal to the optimal water resources allocation and economic development (Gao *et al.* 2016; Rogers *et al.* 2020).

However, such large-scale water conservancy projects also brought a series of hydro-environmental impacts on the downstream river channels (Zhou 2020; Sayed *et al.* 2021). As the largest tributary of the Yangtze River, the middle and lower Han River suffered from the hydrological impact resulting from the TGR and SNWTP. The middle and lower Han River has been seriously polluted for six consecutive years after the middle route of SNWTP operation, which resulted in the frequent occurrence of river eutrophication (Jing *et al.* 2019). Diatom blooms frequently broken out after the SNWTP operation despite adopting some mitigation measures, e.g., the Yangtze-Han Water Diversion Project (YHWDP) (Zhu *et al.* 2008). The current river regulation problem urgently needs to be solved is how to alleviate the adverse influences resulting from the TGR and the middle route of SNWTP in the lower Han River and maintain hydro-environmental health. In recent years, although the effect of the middle route of SNWTP on the middle and lower Han River was studied from different perspectives, little quantitative research was conducted to examine the TGR impact on the hydro-environmental processes of downstream tributaries (Wilson *et al.* 2017; Zhou *et al.* 2017; Zhang *et al.* 2018; Du *et al.* 2019; Zhuang *et al.* 2019). Besides, the former research on the middle route of SNWTP impact on the hydro-environment of the middle and lower Han River is mainly based on assuming or planning conditions because the middle route of SNWTP was just put into operation since 2014 (Zhu *et al.* 2008; Jing *et al.* 2019). However, the actual hydrological processes of the Han River are bound to be different from the planning situations. The numerous environmental factors' variation characteristics and response mechanisms in the middle and lower Han River before and after the scientific identification of water diversion need to be further studied.

This study focused on the hydro-environmental response of the middle and lower Han River to the TGR and the middle route of the SNWTP operation. The research aimed to clarify two key questions: (1) What changes have occurred in hydrology, water quality in the middle and lower Han River before and after the TGR, and the middle route of SNWTP operation? (2) How to identify and quantify the different influences that resulted from both projects, which provides a basis for revealing the causes of the frequent blooms in the lower Han River. Based on a hydro-environmental model and the statistical methods, hydrological changes and the water quality variations under the flow regulation of the TGR and the middle route of SNWTP operation were examined, and quantitative evaluation was conducted in this study. The scientific identification of the formation and evolution mechanism of the hydro-environmental changes in the middle and lower Han River was carried out, which provided an important scientific basis for maintaining the sustainable management of water resources and water environmental health in the Han River basin.

2. MATERIALS AND METHODS

2.1. Study area

The stretch of the Han River falls between 106°~114°E, 30°~34°N, the drainage area is 15.1×10^4 km^2, and the total length of the mainstream is 1,577 km. As shown in Figure 1, Danjiangkou and Zhongxiang divide the mainstream into the upper, middle, and lower reaches. The middle and lower Han River is about 580 km long and flows through Laohekou, Xiangyang, Zhongxiang, Qianjiang, Xiantao, Hanchuan, and Wuhan cities. As the largest tributary of the Yangtze River, the Han River connects its channels with the Yangtze River in Wuhan city. The climate in the Han River basin is mild and humid, and the rainfall is abundant. The multi-year average temperature is about 15~17 °C.

There are four dams (Danjiangkou, Wangfuzhou, Cuijiaying, and Xinglong) along the middle and lower Han River, constituting the cascade reservoirs system. The dams are typical low-head run-of-river hydroelectric plants except for the Danjiangkou dam. The middle route of SNWTP is an enormous inter-basin water transfer project, which significantly contributed to the optimization of China's water resource allocation. The middle route of SNWTP draws water from the Danjiangkou Reservoir (DJKR) in the upper reaches of the Han River. Since its gate of the middle route of SNWTP officially opened on December 12, 2014, it has transferred water up to 2.55×10^{10} m^3 and greatly alleviated the water shortage in northern China.

The TGR is located 603 km upstream of the Han River estuary (Figure 1), which was put into operation in 2003. Significant hydrological changes occurred after the TGR operation during its flow regulation periods, which has an appreciable impact on the downstream flow processes and stage variation along the channels. The variations in the water level at the conference area of the two rivers also affect the outflow from the Han River. Therefore, this study focused on the coupled effects of TGR and the middle route of SNWTP on the hydro-environment along the lower Han River.

2.2. Data source

The research data include topographical, hydrological, meteorological, and water quality datasets. The digital elevation model data was provided by the Bureau of Hydrology, Changjiang Water Resources Commission. The hydrological data were obtained from the hydrological yearbook of the Institute of Geographical Sciences and Natural Resources Research of the Chinese Academy of Sciences. The daily monitoring data were collected, including the inflow and outflow of the Danjiangkou Reservoir, series of gauged flow and stage at the Huangjiagang, Xiangyang, Huangzhuang, Xiantao, and Hanchuan

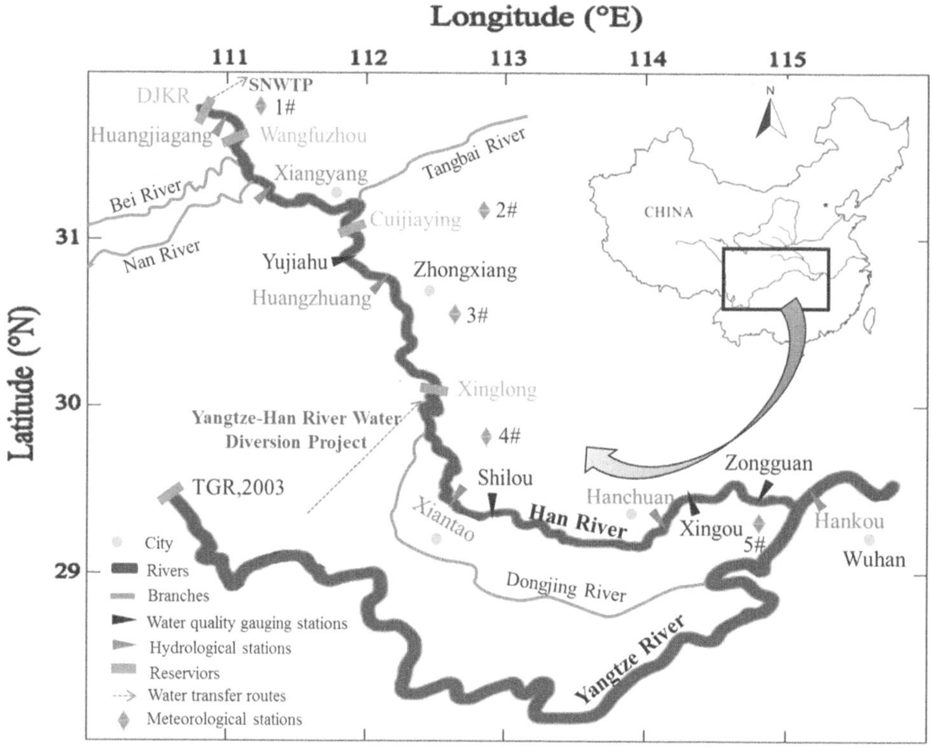

Figure 1 | Overview of the study area.

hydrological stations in the middle and lower Han River and the Hankou station in the Yangtze River. The hydrological data cover the periods from 1996 to 2019.

The meteorological data is downloaded from the China Meteorological Data Network in the middle and lower Han River, including the gauged data at the Laohekou, Zaoyang, Zhongxiang, Tianmen, and Wuhan stations (Figure 1). The time series of meteorological data are consistent with the hydrological data. Besides, the monthly averaged water quality data from 1996 to 2019 at the Shilou, Yujiahu, and Zongguan stations were provided by the Water Environment Monitoring Center of the Yangtze River Basin. The monitoring indicators at these stations include water temperature, ammonia nitrogen (NH_3-N), total phosphorus (TP), biochemical oxygen demand (BOD_5), and pH. Besides, the locations and the waste flow data of the sewage outfalls and water intakes along the middle and lower Han River were collected from the Hubei Environmental Protection Bureau. The data of socioeconomic indicators used in this study was collected from the Hubei Statistical Yearbooks. All the collected data are consistent, and there is no missing data in the time series.

2.3. Research methods

2.3.1. Hydro-environmental modeling for the lower Han River

In this study, a hydro-environmental model was set up for the middle and lower Han River based on the integrated simulation modules developed by the Danish Hydraulic Institute. The framework and modules of the model are shown in Figure 2. The coupled model includes hydrodynamic module and environmental module, and the governing equations are shown as follows.

1D Saint-Venant equations.

Continuity equation:

$$\frac{\partial A}{\partial t} + \frac{\partial Q}{\partial S} = q \tag{1}$$

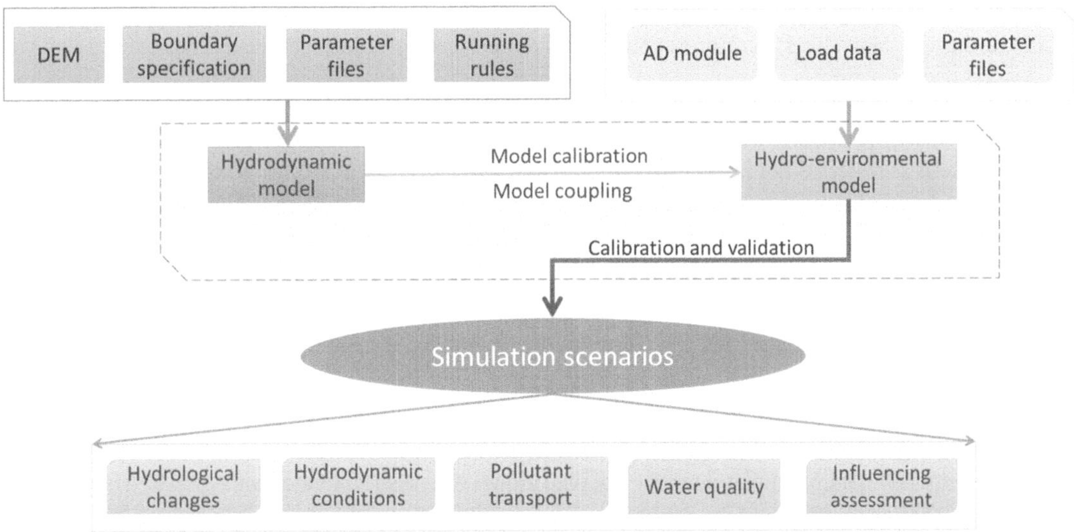

Figure 2 | The modeling framework for the hydro-environmental variation assessment in the middle and lower Han River basin.

Motion equation:

$$\frac{\partial Q}{\partial t}+\frac{\partial (Q^2/A)}{\partial S}=-gA\frac{\partial z}{\partial S}-gA\frac{Q^2}{K^2} \tag{2}$$

where S is the distance between the cross-sections; t is the time; g is the acceleration of gravity; A is the cross-sectional area of the river channels; Q is the flow discharge through the cross-sections; q is the lateral flow per unit width, including inflow and outflow of tributaries and water transfer projects, water consumption for industries and agriculture, the precipitation and evaporation, and the sewage flow; z is the water level; $K=1/nAR^{2/3}$ is the flow modulus, in which R is the hydraulic radius, and n is the Manning resistance coefficient.

Convection-diffusion equation. The convection-diffusion equation was applied to describe the mass transfer in the hydro-environmental model:

$$\frac{\partial \phi}{\partial t}+\frac{\partial (u\phi)}{\partial x}=\frac{\partial}{\partial x}\left(k\frac{\partial \phi}{\partial x}\right)-\eta\phi+S(\phi) \tag{3}$$

where x is the longitudinal coordinate component; u is the mean flow velocity of sections; ϕ is the pollutant concentration; k is the diffusion coefficient; η is the decay rate of pollutants; $S(\phi)$ is the sink/source term of contaminants, including pollutants from the point sources and non-point sources.

Boundary condition treatment. In this study, flow and mass transport boundary conditions were specified in this hydro-environmental model. The inflow of the Danjiangkou reservoir was the upstream flow boundary condition for the coupled model, and the stage-discharge relation curve at the Hankou station was specified as the downstream model condition. The lateral inflow and outflow of branches and the agricultural and industrial water demand were calculated as the source or sink terms of the continuity equation. Moreover, the cascade reservoirs downstream of the Danjiangkou Reservoir are low-head run-of-river hydropower plants, and their scheduling rules are not complex. The reservoir stage will be kept at the normal pool level when the upstream inflow is less than the designed flow.

On the contrary, the reservoir will discharge according to its discharge capacity curve when the inflow is larger than the designed flow. In this study, the dispatching rules of cascade reservoirs are adopted during the hydro-environmental model setup. Besides, the load boundary conditions were specified at the sewage outfall and water intakes based on the collected data. The non-point loads from the subbasins were estimated using the tool of Mike Load Calculator, which was also specified in the coupled model (Zhu *et al.* 2008). The detailed information on the boundary conditions for flow and pollution load along the middle and lower Han River were presented in the Supplementary Material (Text S1).

Model calibration and validation. The middle and lower channel of the Han River was divided into four reaches in the coupled model by three reservoirs (Wangfuzhou, Cuijiaying, and Xinglong) with 316 sections. The largest section distance is 3.51 km, and the average distance is 1.83 km. The hydrological calibration of the model was conducted using the hydrological processes from April 1, 2013 to December 31, 2015, and its validation was carried out by the gauged hydrological data from January 1, 2016 to December 31, 2017. As shown in Supplementary Material, Table S1, the calibrated Manning's roughness coefficient (n) ranged from 0.018 to 0.053. The calibration and validation results are shown in Supplementary Material, Figure S1. The simulation results were evaluated using the Nash-Sutcliffe efficiency coefficients (NSE), and the normalized root mean squared error (NRMSE) (Zhang *et al.* 2018). The NSE at the gauging stations is larger than 0.92, and RMSE is smaller than 0.16, implying the coupled model can be applied to predict and evaluate the hydrological changes accurately under the influences of inter-basin water resource development.

The parameters for the water quality simulation of the coupled model were also calibrated and validated based on the observed data in 2006 and 2007. The results indicate that the longitudinal dispersion coefficient k ranges from 0.035 to 0.050 m^2/s. The detailed information for the model calibration results is shown in Table S1. Supplementary Material, Figure S2 presents the comparison results of water quality indicators between the observed and the simulated data. The statistical comparison results indicate that the average relative errors of NH_3-N, TP, and BOD_5 are 23.5%, 20.1%, and 17.9%, respectively. Therefore, the coupled model can be used to simulate the actual hydro-environmental processes within the acceptable error range.

Simulation scenarios for hydro-environmental evaluation. Based on the 1D hydro-environmental model and the collected data, as shown in Table 1, three simulation scenarios for the hydro-environmental evaluation of the middle and lower Han River were conducted in this study. Scenario 1 was applied to simulate the hydro-environmental processes without the water transfer projects (SNWTP, YHWTP) and TGR. Scenario 2 was used to reveal the backwater effect in the lower Han River. Moreover, Scenario 3 was applied to examine the coupled influences that resulted from the water transfer projects and TGR.

2.3.2. Hydro-environmental changes description

Spearman correlation analysis. In this study, the Spearman correlation analysis was applied to examine the interannual variation of the hydro-environmental regime in the lower Han River. The Spearman correlation analysis was one of the common nonparametric test methods, which can quantify the relationship between the variables (Lobo & Guntur 2018). The following formula can calculate the Spearman rank correlation coefficients of datasets:

$$r_s = 1 - \frac{6\sum d_i^2}{n(n^2 - 1)} \tag{4}$$

in which, r_s is the Spearman rank correlation coefficient, $-1 < r_s < 1$; d_i are the rank differences of a couple of samples within the computing variables; n is the sample size of computing variables. The significance test was examined by comparing the r_s and its critical value.

When the sample size is less than 50, the critical value can be obtained to check the table enclosed to the Spearman rank correlation coefficients. Moreover, when the sample size is more than 50, the critical value can be calculated by the

Table 1 | Simulation scenarios for the hydro-environmental evaluation in the lower Han River

Scenarios	Time interval	TGR	SNWTP + YHWTP
1	1996–2019	X	X
2	1996–2019	✓	X
3	1996–2019	✓	✓

Note: ✓ indicates involving the engineering influences, while X presents having no engineering impacts.

t-test as follows:

$$t = r_s\sqrt{\frac{n-2}{1-r_s^2}} \tag{5}$$

where n is the sample size of computing variables. The significance test can be conducted by comparing the calculated $|t|$ and the critical value t with the $n-2$ degree of freedom under the confidence level of α.

Flow distribution calculation. The coefficient of variation (C_v) was applied to reflect the uneven flow distribution in the lower Han River with and without the engineering influences, and its calculation formula was shown as follows:

$$C_v = \partial/\bar{F} \tag{6}$$

$$\partial = \sqrt{\sum_{i=1}^{12}(F_i - \bar{F})^2} \tag{7}$$

$$\bar{F} = \frac{1}{12}\sum_{i=1}^{12} F_i \tag{8}$$

where ∂ is the coefficient of unevenness; $\bar{F}$ is the annual average flow; F_i is the monthly average flow. It can be seen from Equations (6)–(8) that the larger value of C_v, the more uneven flow distribution occurred during the whole year, and vice versa.

Flow variation range (C_m) is the ratio of the maximum and minimum values of the monthly average flow, reflecting the changing magnitude of the flow during the whole year.

$$C_m = R_{\max}/R_{\min} \tag{9}$$

where $R_{\max}$ is the maximum monthly average flow; $R_{\min}$ is the minimum monthly average flow.

2.3.3. Water quality evaluation

In this study, the integrated evaluation index was applied to evaluate the water quality of the Han River, which examines the water quality by comparing the extreme value or average value of the evaluation factors with the highest allowable value. The equations were shown as follows (Ministry of Ecology & Environment 2018):

$$p_i = \frac{c_i}{c_0} \tag{10}$$

$$P = \frac{1}{n_i}\sum_{i=1}^{n_i} p_i \tag{11}$$

where p_i is the ratio of a pollution index; c_i is the maximum measured concentration value (the minimum value of dissolved oxygen) of a pollution index; c_0 is the highest allowable value of the same index, P is the integrated evaluation index, and n_i is the number of pollutant components. The larger the P-value, the worse water quality there is in the Han River. The water quality grade is determined by calculating the integrated evaluation index P as follows: (i) excellent ($P < 1$), (ii) good ($1 \leq P < 2$), (iii) marginal ($2 \leq P < 3$), (iv) Poor ($3 \leq P \leq 5$), and (v) very poor ($P > 5$) (Pesce & Wunderlin 2000; Kannel *et al.* 2007). To better evaluate the nutrient status of water columns and the concentration of all biochemically degradable oxygen-consuming organic matter in the middle and lower Han River, the water quality indicators NH_3-N, TP, and BOD_5 were selected in this study.

2.3.4. Nutrient assimilative capacity calculation

Nutrient assimilative capacity evaluation refers to the pollutant amount that the water can accommodate or the ability to purify and maintain ecological balance without affecting the normal use of water. The pollutants in the small and medium rivers are easily mixed across the sections because of their large ratio between the river length, width, and depth. Therefore, 1D hydro-environmental model can be applied to simulate the environmental processes and predict the variations in nutrient assimilative capacity because the coupled model meets the steady-state requirements. Based on the convection equation of the 1D hydro-environmental model, the nutrient assimilative capacity can be calculated using the following Equation (12).

$$C_s = \frac{QC_0 + W_1}{Q + q} \tag{12}$$

in which, W_1 is the nutrient assimilative capacity for kinds of pollutants within an objective river reach (g/s); Q is the discharge for the designed flow level (m^3/s); q is the sewage discharge (m^3/s); C_s is the pollutant concentration after mixing in the computing element, and C_0 is the pollutant concentration in the upstream section of the computing element. In this study, the flow ratio, $m = Q/(Q + q)$, was introduced to Equation (12). Thus the nutrient assimilative capacity for the computing element was deduced as follows:

$$W_1 = (C_s/m - C_0) \times Q \tag{13}$$

3. RESULTS

3.1. Statistical characteristics of the hydrological changes

Figure 3 presents the flow duration curve at the Huangjiagang station. If the flow of less than 800 m^3/s was considered the low flow, the low flow duration significantly increased after water diversion, which extended from 3.2–5.1 months to 7.0–10.1 months. As shown in Table 2, the average annual flow at the Xiantao station increased from 843.67 m^3/s to 874.62 m^3/s after the TGR operation, while it decreased to 775.31 m^3/s after water diversion. In particular, there was much more distinct flow alteration during the flood seasons than the dry seasons. It should be noted that the multi-year average flow of scenario 3 decreased both in flood seasons and dry seasons comparing to Scenarios 1 & 2. Spearman correlation analysis was carried out based on the monthly flow of the three scenarios. The test results showed that the $|T|$ was less than the critical value in Scenarios 1 & 2, which indicated no obvious changing trend in flow series during the two periods. In particular, although there were evident flow variations during the dry seasons and flood seasons after the TGR and SNWTP operation, the Spearman test results did not show a distinct changing trend within the statistical periods of the three scenarios.

The calculation results of flow distribution before and after the water diversion are shown in Table 3. There was a significant increase in the coefficient of variation and flow variation range of Scenario 3, which indicated the much more uneven flow distribution occurred after the water diversion. Compared with Scenario 1, there was no significant increase in the hydrologic alteration indicators of Scenario 2, which indicated that the TGR operation exerted slight influences on the flow distribution in the lower Han River.

Moreover, the water level at the estuary of the Han River is the vital outflow boundary condition, which greatly contributed to the outflow processes variation of the lower Han River. The stage variation at the Hankou station can reflect the estuary outflow boundary changes, which are presented in Supplementary Material, Figure S3. There was a significant variation in the water level at the Hankou station after the TGR operation. The monthly average water level decreased obviously from April to December, especially stage decrease during October and November because of TGR impoundment. The maximum stage decrease was up to 2.34 m. On the contrary, the water level increased slightly during the water compensation from the TGR (from January to March), and the average stage increase was 0.78 m.

3.2. Performances assessment of water quality variables

Based on the simulation results of the hydro-environmental model, the changes in water quality indicators, NH_3-N, TP, and BOD_5, were presented in Figure 4. The environmental processes at the Yujiahu, Shilou, and Zongguan stations reflected the spatiotemporal variations in water quality along the Han River channels. As shown in Figure 4, there was a distinctly rising trend of Scenario 3 in NH_3-N concentration at the Yujiahu, Shilou, and Zongguan stations compared with Scenarios 1 and 2. The largest monthly average concentration increase at the Yujiahu station was up to 0.109 mg/l. Compared with Scenario 1, the concentration variations in the NH_3-N of Scenario 2 presented a different trend during the dry and flood seasons. In Scenario 2, there was a slight increase in the NH_3-N concentration during the dry seasons (December to the following February),

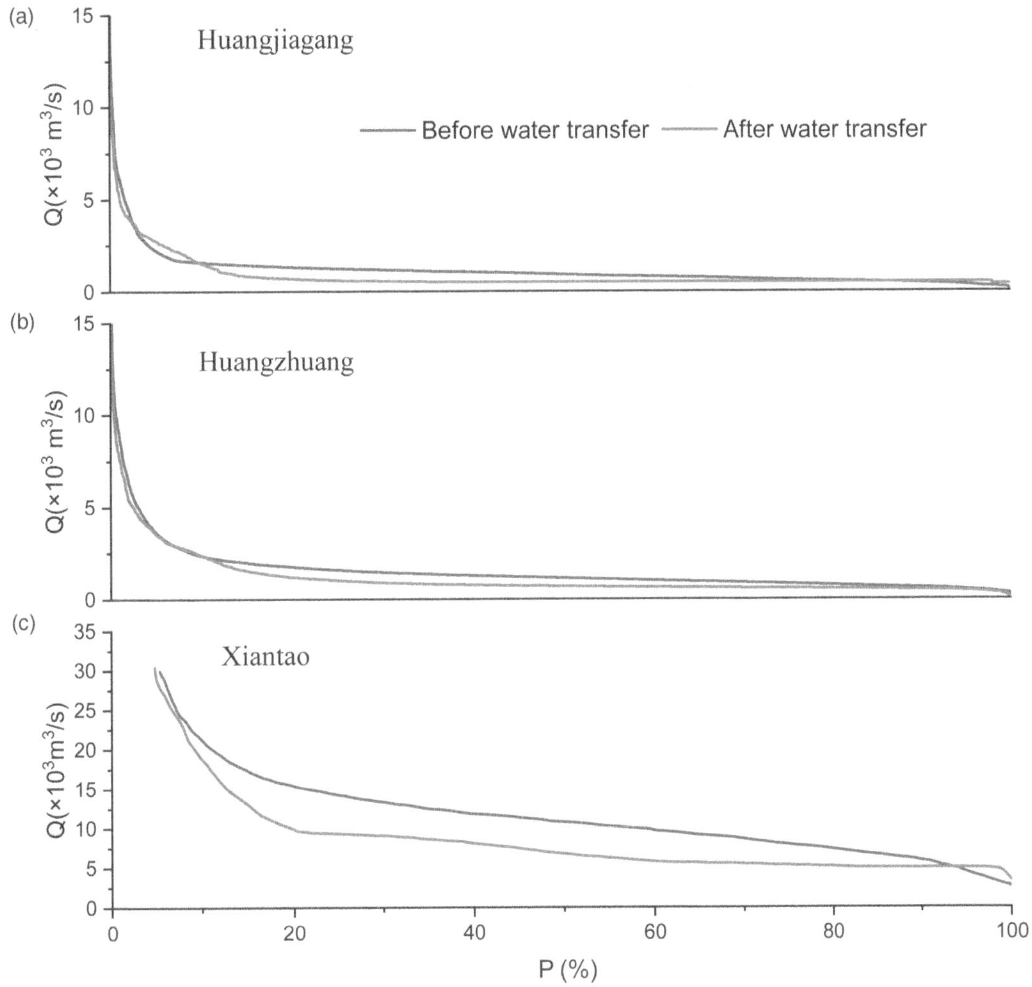

Figure 3 | Flow duration curve at the Huangjiagang station.

Table 2 | Statistical results of the flow series and their Spearman test results at Xiantao station

	Multi-year average flow/(m^3/s)			**Spearman |T|**		
Periods	**Scenario No.1**	**Scenario No.2**	**Scenario No.3**	**Scenario No.1**	**Scenario No.2**	**Scenario No.3**
Annual	843.67	874.62	775.31	0.143	0.081	0.800
Dry seasons	633.31	669.44	604.76	0.119	0.091	0.860
Flood seasons	1,051.31	1,269.79	915.55	0.333	0.100	0.800

Note: Spearman correlation analysis was conducted at the significance level of 0.05. The critical values used for the Spearman test in scenarios No.1-3 were 0.786, 0.587, and 0.900, respectively. The dry seasons involved here were from December to the following March, while the flood seasons were from June to September.

Table 3 | Variation in annual runoff distribution at the gauging stations

	Huangjiagang			**Xiantao**		
Items	**Scenario No.1**	**Scenario No.2**	**Scenario No.3**	**Scenario No.1**	**Scenario No.2**	**Scenario No.3**
Coefficient of variation	1.64	1.64	2.53	1.55	1.69	2.41
Flow variation range	4.35	4.35	6.55	4.21	4.47	6.17

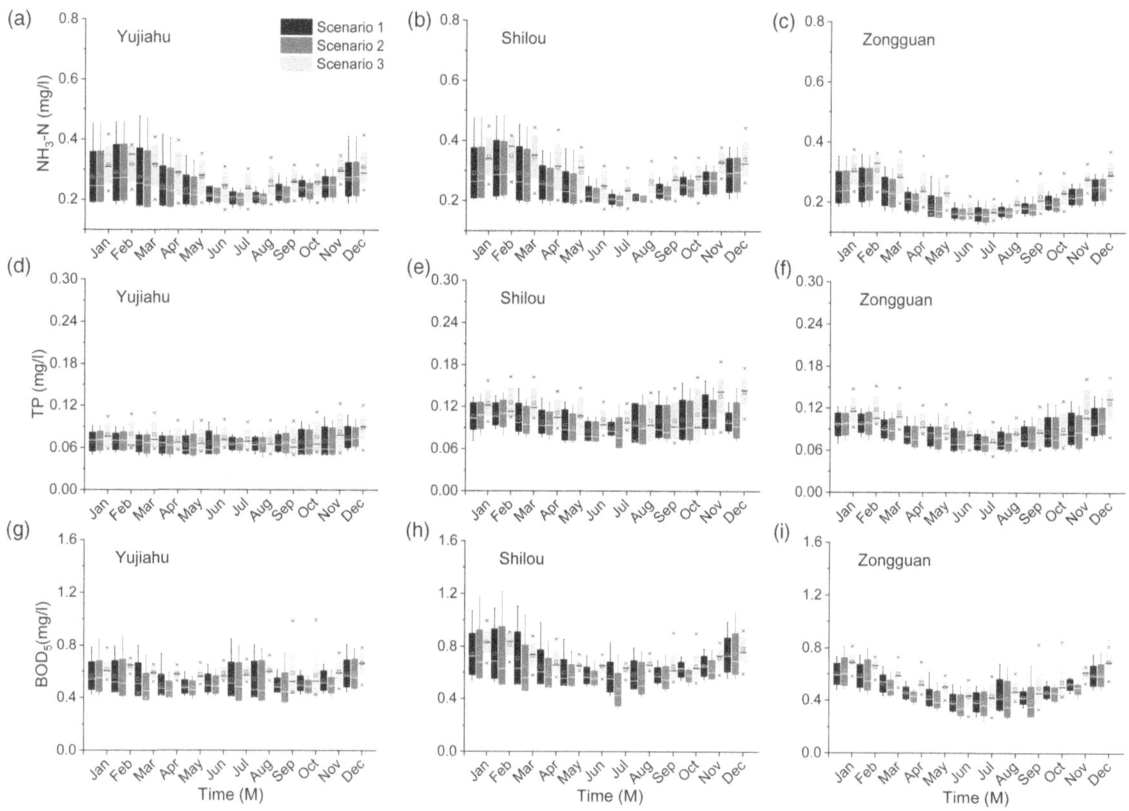

Figure 4 | Boxplots of hydro-environmental indices at the gauging stations.

while it decreased to a different extent at the gauging stations during the flood seasons. The maximum NH_3-N concentration decrease occurred at the Zongguan station by 0.06 mg/l during August. In terms of spatial variation along the channels, the closer to the estuary of the Han River, the more significant concentration variations occurred at the gauging stations. The monthly average NH_3-N concentration increase of Scenario 2 was 0.048 mg/l at the Zongguan station than that of Scenario 1, which increased by 12.6% in contrast to that at the Yujiahu station.

There were similar changing trends of TP, BOD_5 at the gauging stations. In contrast to Scenario 1, the TP and BOD_5 concentrations variated to a different extent in Scenarios 2 & 3. In particular, there was a significant concentration increase in TP and BOD_5 of Scenario 3. Compared with Scenario 1, the largest concentration increases in TP and BOD_5 of Scenario 3 at the gauging stations were 0.04 mg/l and 0.15 mg/l, respectively. In addition, the monthly average concentration of TP and BOD_5 of Scenario 2 decreased by 0.005 mg/l and 0.089 mg/l than the ones of Scenario 1 during the flood seasons. On the contrary, the monthly average TP and BOD_5 concentration of Scenario 2 increased by 0.002 mg/l and 0.037 mg/l than the ones of Scenario 1 during the dry seasons.

Supplementary Material, Figure S4 shows the changing trend of the integrated evaluation index along the middle and lower Han River. There was an obvious increase in the integrated evaluation index of Scenario 3 at the gauging stations. After the TGR and SNWTP were implemented, the integrated evaluation index variated to different degrees at the Yujiahu, Shilou, and Zongguan stations. In particular, there was not a good consistency in the variations of integrated evaluation index P in Scenario 2. The largest P-decrease of Scenario 2 was 0.021, 0.035, and 0.041 than Scenario 1 during the periods from March to November, while the largest P-increase were 0.018, 0.020, and 0.022 during the dry seasons (December to the following February). Compared to the increase of integrated evaluation index P between the Shilou and Zongguan stations, the more obvious P-increase occurred at the estuarine Zongguan station of the Han River. It should be noted that the integrated evaluation index was still less than 1.0, which indicated that it was not larger than the highest allowable value along the middle and lower Han River.

The integrated evaluation index was examined by the Spearman test, and the results are shown in Table 4. The critical values of the Spearman test in scenarios 1–3 at the significance level of 0.05 are 0.786, 0.587, and 0.900, respectively. The

Table 4 | Spearman test results of the integrated evaluation index at gauging stations

Scenario No.	Periods	Yujiahu	Shilou	Zongguan
1	1996–2019	0.007	−0.182	−0.070
2	1996–2019	−0.503	−0.503	−0.287
3	1996–2019	0.343	0.168	0.021

integrated evaluation index at the gauging stations is less than the critical values. Therefore, there was no obvious increase or decrease trend in the integrated evaluation index at the gauging stations. Moreover, the SNWTP significantly influenced pollutants duration along the middle and lower channels when it was put into operation (Figure 5). In particular, the NH_3-N concentration of Scenario 3 at the stations was larger than 0.30 mg/l during more than 50% of the statistical time. Also, the changing processes of TP and BOD_5 had a similar trend with that of NH_3-N. The duration of higher pollutant concentration was drastically extended in Scenario 3.

3.3. Variations in nutrient assimilative capacity

The hydrological changes resulted in the variations in water quality indicators and influenced the nutrient assimilative capacity along the middle and lower Han River. The spatiotemporal variations in nutrient assimilative capacity are shown in Figure 6. Although the water quality indicators fluctuated, the decreasing trend was evident after the SNWTP operation. The more runoff flowed through the channels, the larger the nutrient assimilative capacity there was along the Han River, e.g., the nutrient assimilative capacity increased significantly during the rainy periods from 2011 to 2013.

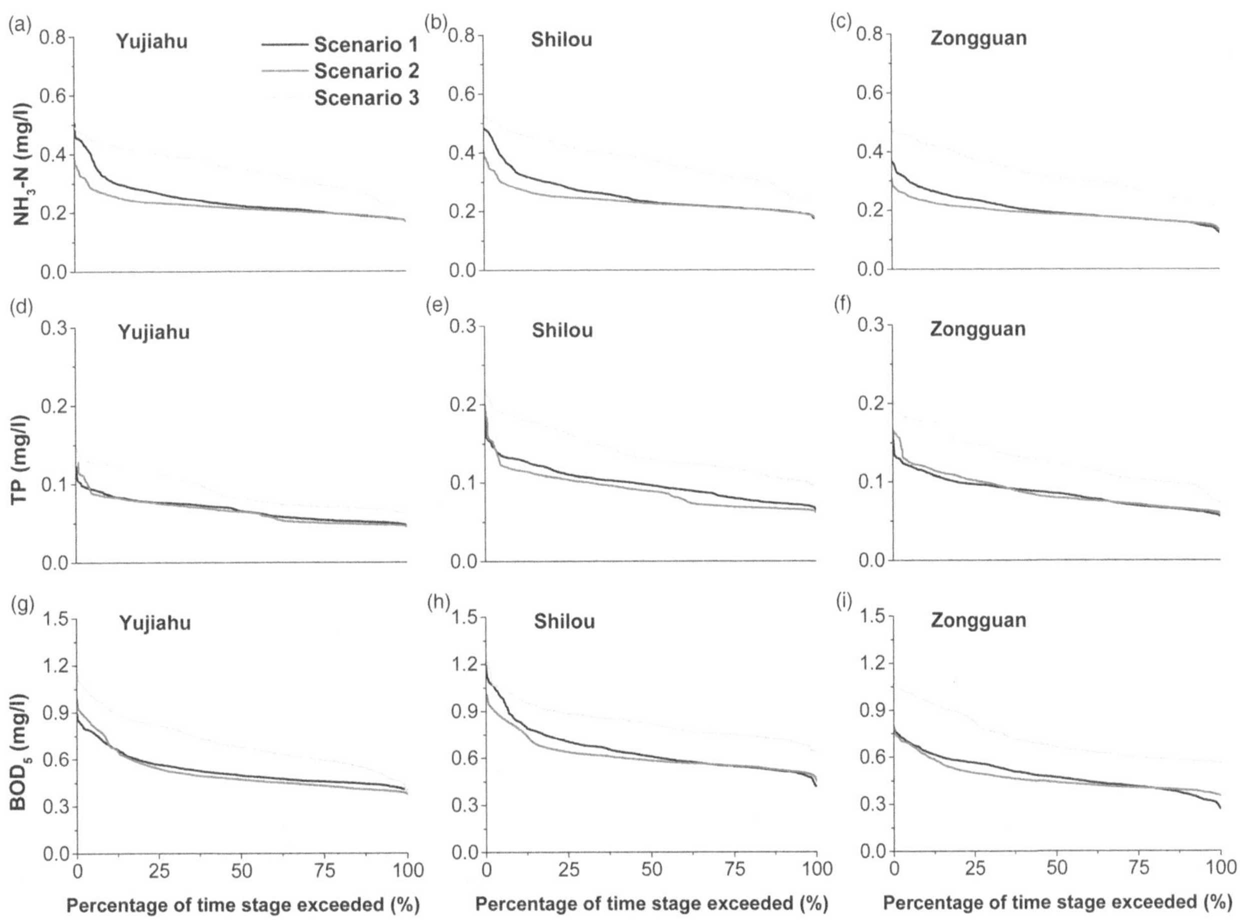

Figure 5 | Pollutants duration curves during the different periods at the gauging stations.

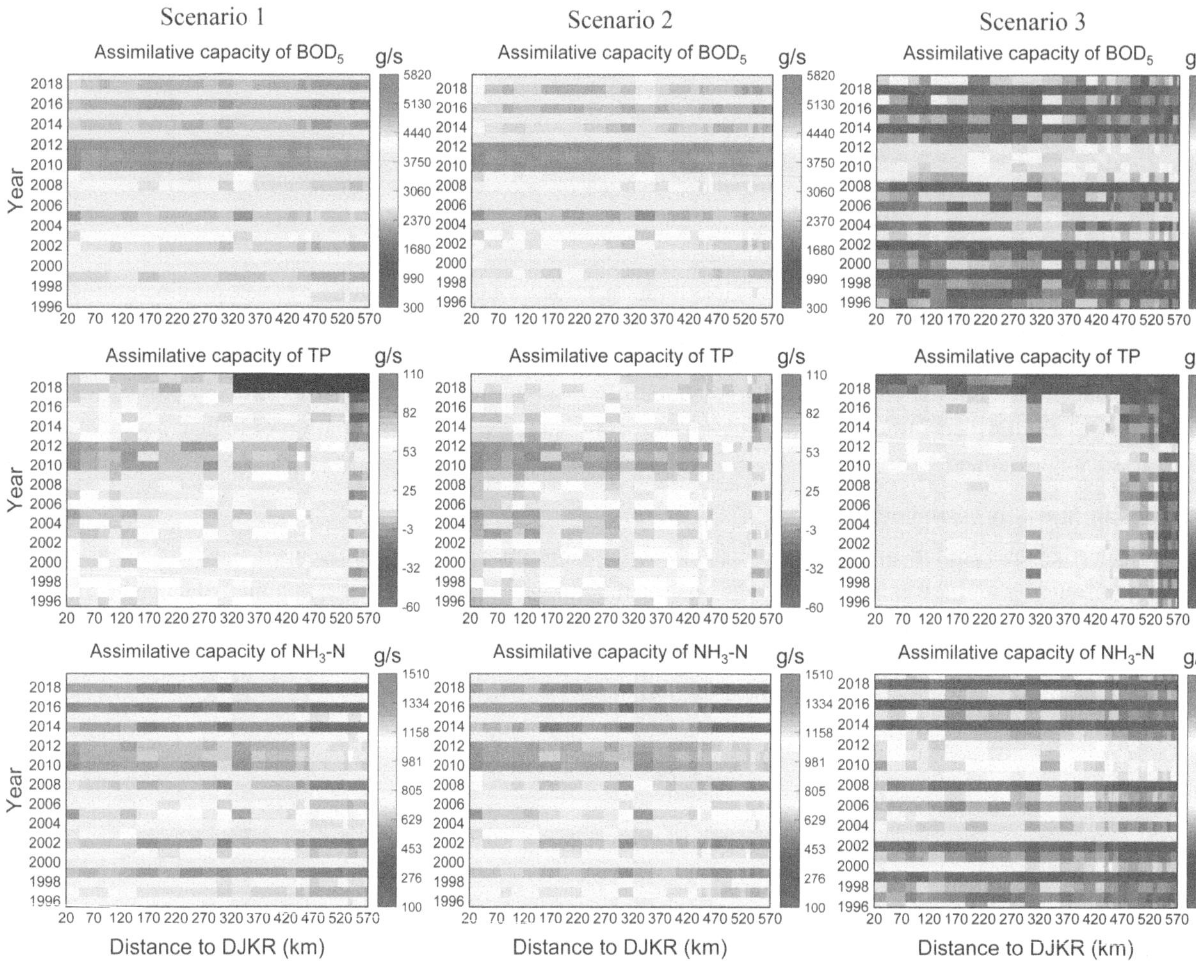

Figure 6 | Variations in nutrient assimilative capacity along the middle and lower Han River.

The average allowable capacity of NH_3-N for the middle and lower Han River was 853.85 g/s, 870.69 g/s, and 586.27 g/s in Scenarios 1, 2, and 3, respectively. As shown in Figure 6, the average allowable capacity of NH_3-N was much larger than that of TP. The TP was evident out of the limit downstream of the Xiantao city. In particular, the closer to the estuary of the Han River, the larger decrease occurred in nutrient assimilative capacity. Also, there was a similar spatio-temporal variation in the BOD_5 along the lower Han River. With the increase of pollutants, it needed more oxygen to decompose the organic matter. Therefore, the available capacity for the BOD_5 along the middle and lower Han River decreased significantly.

4. DISCUSSION

As shown in the results, the water environment of the Han River is deteriorating under the influences of inter-basin water resource development. The hydrological changes along the middle and lower Han River significantly contributed to the water environmental variations. The runoff decreased substantially after the SNWTP operation, directly increasing pollutant concentration and rapidly decreasing nutrient assimilative capacity (Figures 4–6).

The flow decrease in the Han River and the backwater effect of the water surface of the Yangtze River increased the estuarine water level of the Han River, which led to the hydraulic gradient decrease and velocity reduction along the lower Han River to a different extent. The monthly averaged flow velocity at the Yujiahu, Shilou, and Zongguan stations was presented in Supplementary Material, Figure S5. The flow velocity during the dry seasons decreased slightly after the TGR and SNWTP operation, and the flow velocity larger than 1.5 m/s rarely occurred along the lower channels. In particular, the monthly average flow velocity decreased by 0.1 m/s during the dry seasons under the coupled influences of the TGR and SNWTP (Supplementary Material, Figure S5). Due to the evident runoff reduction and flow velocity decrease, the pollutant dispersion becomes much slower. Therefore, water self-purification capacity decreased and the pollutants accumulated in the middle and lower Han River.

However, there was no consistency in the flow variations all year because of the different floods encountered between the two rivers. Compared with Scenario 1, the TGR indeed contributed to a certain degree to the estuarine level decrease and flow velocity increase along the lower Han River during the TGR impounding periods (September and October). There was a flow velocity increase by 0.13–0.52 m^3/s in Scenario 2 during the TGR impounding periods. However, the flow velocity increase during the TGR impoundment was still unable to improve the water environment of the middle and lower Han River. Because the runoff decrease of the middle and lower Han River drastically reduced the water level after the SNWTP operation, there was no extremely significant flow velocity increase at all. In particular, the TGR impoundment impact lasts no more than 2.5 months annually. Although the flow in the Han River increased after the YHWDP was put into operation, the low flow duration had no remarkable changes (Zhang *et al.* 2018). Therefore, the YHWDP can improve the water quality after its operation but will not completely offset the adverse influences resulting from the SNWTP. The runoff decrease is the primary cause of water quality deterioration along the middle and lower Han River.

The concentration increase in NH_3-N, TP, and BOD_5 deteriorated the water quality in the lower Han River, which significantly reduced the nutrient assimilative capacity. Consequently, several algal bloom events occurred in recent years, which posed a great threat to water safety in the Han River basin (Xin *et al.* 2020). To deal with the water quality degradation and the series of negative influences resulted from the project operation on the hydro-environment, the outflow into the middle and lower channels of Han River should be increased to offset the flow decrease after the SNWTP operation. Moreover, pollutant emissions should be strictly limited, and wastewater treatment should also be taken seriously in the Han River basin. Last but not least, optimizing water resource scheduling of the inter-basin water resource may also be an effective measure to deal with the deterioration of water quality under the limited water quantity.

5. CONCLUSION

The SNWTP and TGR, as the important hydraulic engineering systems to optimize water distribution across watersheds, also had significant adverse effects on downstream river health. In this study, the hydro-environmental modeling and the middle and lower Han River data analysis were applied to evaluate the hydrological changes and water-quality variations resulting from the inter-basin water resource regulation. The lower Han River is under the combined effects of the hydrological variations in the Han River and the Yangtze River. After the SNWTP operation, the low flow duration was extended from 3.2–5.1 months to 7.0–10.1 months.

On the other hand, the water level of the Han River estuary, as one of the most important outflow boundary conditions, had remarkable variations after the TGR operation. There was an obvious stage decrease during the TGR impoundment, while the estuarine stage of the Han River increased slightly when the TGR began to supplement water to the downstream channels. The water quality of the middle and lower Han River deteriorated under the influences of the inter-basin water resource regulation of the SNWTP. The concentration increase in NH_3-N, TP, and the BOD_5 resulted in an obvious increase in the integrated evaluation index at the gauging stations. Consequently, the allowable nutrient assimilative capacity decreased significantly. Effective measures should be taken to improve the hydro-environment along the middle and lower Han River to reduce pollutant emission and optimize the reservoir operation.

ACKNOWLEDGEMENTS

This work was supported by the National Key Research and Development Plan (2016YFC0400901) and the National Natural Science Foundation of China (51509273).

DECLARATION OF COMPETING INTEREST

The authors have declared no conflict of interest.

DATA AVAILABILITY STATEMENT

All relevant data are included in the paper or its Supplementary Information.

REFERENCES

Bakker, K. 2012 Water security: research challenges and opportunities. *Science* **337** (6097), 914–915.

Bonacci, O. & Oskorus, D. 2010 The changes in the lower Drava River water level, discharge and suspended sediment regime. *Environmental Earth Sciences* **59** (8), 1661–1670.

Bond, N. R., Lake, P. S. & Arthington, A. H. 2008 The impacts of drought on freshwater ecosystems: an Australian perspective. *Hydrobiologia* **600** (1), 3–16.

Cheng, K., Wei, S., Fu, Q., Pei, W. & Li, T. 2019 Adaptive management of water resources based on an advanced entropy method to quantify agent information. *Journal of Hydroinformatics* **21** (3), 381–396.

Du, W. Y., Fan, Y. B., Liu, X. J., Park, S. C. & Tang, X. W. 2019 A game-based production operation model for water resource management: an analysis of the South-to-North Water Transfer Project in China. *Journal of Cleaner Production* **228**, 1482–1493.

Gao, Q., Li, Y., Cheng, Q., Yu, M., Hu, B., Wang, Z. & Yu, Z. 2016 Analysis and assessment of the nutrients, biochemical indexes and heavy metals in the Three Gorges Reservoir, China, from 2008 to 2013. *Water Research* **92**, 262–274.

Gurría, A. 2009 Sustainably managing water: challenges and responses. *Water International* **34** (4), 396–401.

Hao, Z., Rallings, A. M., Espinoza, V., Luo, P., Duan, W., Peng, Q., Gao, Y. & Viers, J. H. 2021 Flowing from east to west: a bibliometric analysis of recent advances in environmental flow science in China. *Ecological Indicators* **125**, 107358.

Jing, Z., Xia, J., Zhang, X., Wang, Q. & Ma, X. 2019 Spatial and temporal distribution and variation of water quality in the middle and downstream of Hanjiang River. *Research of Environmental Sciences* **32** (1), 104–115.

Kannel, P. R., Lee, S., Lee, Y., Kanel, S. R. & Khan, S. P. 2007 Application of water quality indices and dissolved oxygen as indicators for river water classification and urban impact assessment. *Environmental Monitoring and Assessment* **132** (1), 93–110.

Koster, W. M., Stuart, I., Tonkin, Z., Dawson, D. & Fanson, B. 2021 Environmental influences on migration patterns and pathways of a threatened potamodromous fish in a regulated lowland river network. *Ecohydrology* **14** (2), e2260.

Liechti, T. C., Matos, J. P., Boillat, J. L. & Schleiss, A. J. 2015 Influence of hydropower development on flow regime in the Zambezi River Basin for different scenarios of environmental flows. *Water Resources Management* **29** (3), 731–747.

Lobo, M. & Guntur, R. D. 2018 Spearman's rank correlation analysis on public perception toward health partnership projects between Indonesia and Australia in East Nusa Tenggara Province. *Journal of Physics Conference* **1116**, 1–6.

Ministry of Ecology and Environment, C. 2018 *Technical Guidelines for Environmental Impact Assessment – Surface Water Environment (HJ2.3-2018)*. China Environmental Science Press, Beijing, China.

O'Connor, J. E., Duda, J. J. & Grant, G. E. 2015 1000 dams down and counting. *Science* **348** (6234), 496–497.

Palmer, M. A. 2010 Beyond infrastructure. *Nature* **467** (7315), 534–535.

Pesce, S. F. & Wunderlin, D. A. 2000 Use of water quality indices to verify the impact of Córdoba City (Argentina) on Suquı´a River. *Water Research* **34** (11), 2915–2926.

Pruden, A., Arabi, M. & Storteboom, H. N. 2012 Correlation between upstream human activities and riverine antibiotic resistance genes. *Environmental Science & Technology* **46** (21), 11541–11549.

Rogers, S., Chen, D., Jiang, H., Rutherfurd, I., Wang, M., Webber, M., Crow-Miller, B., Barnett, J., Finlayson, B. & Jiang, M. 2020 An integrated assessment of China's south – north water transfer project. *Geographical Research* **58** (1), 49–63.

Sayed, E. T., Wilberforce, T., Elsaid, K., Rabaia, M. K. H., Abdelkareem, M. A., Chae, K.-J. & Olabi, A. G. 2021 A critical review on environmental impacts of renewable energy systems and mitigation strategies: wind, hydro, biomass and geothermal. *Science of The Total Environment* **766**, 144505.

Scionti, F., Miguez, M. G., Barbaro, G., De Sousa, M. M., Foti, G. & Canale, C. 2018 Integrated methodology for urban flood risk mitigation in cittanova, Italy. *Journal of Water Resources Planning and Management* **144** (10), 05018013.

Vörösmarty, C. J., McIntyre, P. B., Gessner, M. O., Dudgeon, D., Prusevich, A., Green, P., Glidden, S., Bunn, S. E., Sullivan, C. A., Liermann, C. R. & Davies, P. M. 2010 Global threats to human water security and river biodiversity. *Nature* **467** (7315), 555–561.

Wilson, M. C., Li, X. Y., Ma, Y. J., Smith, A. T. & Wu, J. G. 2017 A review of the economic, social, and environmental impacts of China's south-north water transfer project: a sustainability perspective. *Sustainability* **9** (8), 1–11.

Xin, X. K., Zhang, H., Lei, P., Tang, W. Z., Yin, W., Li, J., Zhong, H. & Li, K. F. 2020 Algal blooms in the middle and lower Han River: characteristics, early warning and prevention. *Science of The Total Environment* **135293**, 1–9.

Yang, J. R., Lv, H., Isabwe, A., Liu, L., Yu, X., Chen, H. & Yang, J. 2017 Disturbance-induced phytoplankton regime shifts and recovery of cyanobacteria dominance in two subtropical reservoirs. *Water Research* **120**, 52–63.

Zhang, J. H., Feng, L. J., Chen, S. J., Huang, T., Chen, L., Wang, D. W., Dai, M. L. & Zhang, D. D. 2018 Impoundment impact of the three gorge reservoir on the hydrological regime in the Lower Han River, China. *Water* **10** (11), 1–17.

Zhou, Y. 2020 Exploring multidecadal changes in climate and reservoir storage for assessing nonstationarity in flood peaks and risks worldwide by an integrated frequency analysis approach. *Water Research* **185**, 116265.

Zhou, Y., Guo, S., Hong, X. & Chang, F. J. 2017 Systematic impact assessment on inter-basin water transfer projects of the Hanjiang River Basin in China – scienceDirect. *Journal of Hydrology* **553**, 584–595.

Zhu, Y. P., Zhang, H. P., Chen, L. & Zhao, J. F. 2008 Influence of the south–north water diversion project and the mitigation projects on the water quality of Han River. *Science of The Total Environment* **406** (1), 57–68.

Zhuang, W., Ying, S. C., Frie, A. L., Wang, Q., Song, J. M., Liu, Y. X., Chen, Q. & Lai, X. Y. 2019 Distribution, pollution status, and source apportionment of trace metals in lake sediments under the influence of the South-to-North Water Transfer Project, China. *Science of The Total Environment* **671**, 108–118.

First received 20 August 2021; accepted in revised form 16 November 2021. Available online 4 December 2021

doi: 10.2166/nh.2021.007

Probabilistic interval estimation of design floods under non-stationary conditions by an integrated approach

Yanlai Zhou [a,b,*], Shenglian Guo[a], Chong-Yu Xu[b], Lihua Xiong[a], Hua Chen[a], Cosmo Ngongondo[c] and Lu Li[d]

[a] State Key Laboratory of Water Resources and Hydropower Engineering Science, Wuhan University, Wuhan 430072, China
[b] Department of Geosciences, University of Oslo, P.O. Box 1047 Blindern, Oslo N-0316, Norway
[c] Department of Geography and Earth Sciences, University of Malawi, Chancellor College, P.O. Box 280, Zomba, Malawi
[d] NORCE Norwegian Research Centre, Bjerknes Centre for Climate Research, Bergen NO-5007, Norway
*Corresponding author. E-mail: yanlai.zhou@whu.edu.cn

YZ, 0000-0002-5447-2420

ABSTRACT

Quantifying the uncertainty of non-stationary flood frequency analysis is very crucial and beneficial for planning and design of water engineering projects, which is fundamentally challenging especially in the presence of high climate variability and reservoir regulation. This study proposed an integrated approach that combined the Generalized Additive Model for Location, Scale and Shape parameters (GAMLSS) method, the Copula function and the Bayesian Uncertainty Processor (BUP) technique to make reliable probabilistic interval estimations of design floods. The reliability and applicability of the proposed approach were assessed by flood datasets collected from two hydrological monitoring stations located in the Hanjiang River of China. The precipitation and the reservoir index were selected as the explanatory variables for modeling the time-varying parameters of marginal and joint distributions using long-term (1954–2018) observed datasets. First, the GAMLSS method was employed to model and fit the time-varying characteristics of parameters in marginal and joint distributions. Second, the Copula function was employed to execute the point estimations of non-stationary design floods. Finally, the BUP technique was employed to perform the interval estimations of design floods based on the point estimations obtained from the Copula function. The results demonstrated that the proposed approach can provide reliable probabilistic interval estimations of design floods meanwhile reducing the uncertainty of non-stationary flood frequency analysis. Consequently, the integrated approach is a promising way to offer an indication on how design values can be estimated in a high-dimensional problem.

Key words: Bayesian Uncertainty Processor (BUP), Copula function, flood frequency analysis, Hanjiang River, non-stationarity

HIGHLIGHTS

- This study proposes an integrated approach to reduce uncertainties of flood frequency analysis.
- The GAMLSS method models time-varying characteristics in marginal and joint distributions.
- The Copula function makes point estimations of non-stationary design floods.
- The BUP creates reliable interval estimations of non-stationary design floods.

1. INTRODUCTION

Floods are among the world's costliest natural disasters, and changing environments have made catastrophic floods more likely. Numerous lessons have been learnt about the failure of hydraulic structures related to changing environments or changing characteristics of annual maximum streamflow series (Hui *et al.* 2018; Ragno *et al.* 2019). The flood frequency analysis results have been widely applied in the planning and design of a dam/reservoir and other civil engineering applications (Li *et al.* 2020). Conventional flood frequency analysis methods are based on the assumption of temporal stationarity that the return period (or occurrence probability) of extreme floods will not alter significantly with time. In the presence of climate and anthropogenic changes, the stationary assumption of hydrological extreme series has been challenged (Cheng & AghaKouchak 2015; Yan *et al.* 2017a, 2021). If flood non-stationarity was not adequately considered, flood risks based on the stationarity assumption would be miscalculated (under-/overestimated) in practice (Li *et al.* 2015; Serago & Vogel 2018).

Therefore, non-stationary frequency analysis has been an important research topic for the ultimate purpose of supporting the planning and design of hydraulic engineering projects.

The statistical characteristics of hydrological series are commonly altered by climate change and/or human activities (Song *et al.* 2020). Such changing characteristics include the changes in statistical parameters and/or the type of probability distributions (Xiong *et al.* 2017; Liu *et al.* 2019; Sun *et al.* 2020; Zhou 2020). In recent years, several methods were used to describe the variation of parameters and weighting coefficients in distributions of hydrological series, such as single-type distributions (Gilroy & McCuen 2012), two-component mixture distributions (Bayazit 2015; Yan *et al.* 2017b) and time-varying moments (Xiong *et al.* 2015a; Yu *et al.* 2018). Additionally, the change point and trend change detection methods such as the Pettitt test (Pettitt 1979; Reeves *et al.* 2007; Ngongondo *et al.* 2020), the Mann–Kendall test (Chebana *et al.* 2013) and the trend-free pre-whitening method (Yue *et al.* 2002) were applied to detect the variability of hydrological series (Villarini & Smith 2010; Şen 2011; Rougé *et al.* 2013). Although the change point and trend change detection methods can be used to detect the hydrological non-stationarity, they cannot quantify the effects of possible physical factors on the non-stationarity of hydrological series (López & Francés 2013; Villarini & Strong 2014; Zhang *et al.* 2015; Wu *et al.* 2017; Liang *et al.* 2018; Su & Chen 2019; Bian *et al.* 2020). The Generalized Additive Model for Location, Scale and Shape parameters (GAMLSS) (Stasinopoulos & Rigby 2007) method provides a higher flexible choice in modeling the time-varying characteristics of distribution parameters, as compared with classical generalized additive models, generalized linear models, generalized linear mixed models and generalized additive mixed models (Villarini *et al.* 2009). Hence, in this study, the GAMLSS method is adopted to model the impacts of possible physical factors on the annual maximum streamflow series.

The joint probability distribution of multiple hydrological variables is very important in the hydraulic planning and design of a river basin (Parent *et al.* 2014; Li *et al.* 2018). The implementation of hydrological frequency analysis is fundamentally challenging for multivariate design variables. As compared with the multivariate functions, such as Gaussian, Student's *t* and Gamma (Chen & Guo 2018; Zhang & Singh 2019), Copula functions have the flexibility in choosing marginal distribution and modeling a non-linear dependence of variables when fitting hydrological data (Kwon & Lall 2016; Sarhadi *et al.* 2016; Chen & Guo 2018; Wei & Song 2018; Tan *et al.* 2021). For instance, Jiang *et al.* (2015) applied Copula functions to make bivariate frequency analysis of low-streamflow series. Fan *et al.* (2016) fused Gaussian mixtures into a Copula function to implement the flood frequency analysis for estimating hydrological risks. Liu *et al.* (2017) integrated the maximum entropy and Copula functions to model the non-linear dependence of multiple hydro-meteorological events. Vinnarasi & Dhanya (2019) adopted a dynamic Copula function to model the non-stationarity in extreme rainfall series. Consequently, in this study, the Copula function is employed to derive the joint distribution of multiple annual maximum streamflow series.

One of the effective techniques to quantify uncertainties in flood frequency analysis is to create probabilistic interval estimations of design floods (Read & Vogel 2015; Yan *et al.* 2019). The combination of point estimation method and probabilistic post-processing technique has been a promising integrated approach for quantifying the impacts of the uncertainties of distribution types and parameters on flood frequency analysis. Probabilistic interval estimation was commonly used to supplement the information provided by point estimations (Bracken *et al.* 2018). The Bayesian Uncertainty Processor (BUP) technique proposed by Krzysztofowicz (1999) is able to quantify the uncertainty associated with flood frequency analysis, owing to its ability to facilitate the development of the interval estimation (e.g. Thorarinsdottir *et al.* 2018; Xu *et al.* 2018; Li & Krafty 2019; Uranchimeg *et al.* 2020). Therefore, it is interesting and important to conduct an in-depth research on the exploration of the BUP post-processing technique for reducing the uncertainty encountered in non-stationary flood frequency analysis.

This study is aimed at acquiring reliable probabilistic interval estimations of non-stationary design floods. The novelty of this study lies in proposing an integrated frequency analysis approach by combining the GAMLSS method, the Copula function and the BUP post-processing technique to reduce the uncertainty associated with flood frequency analysis for the first time. First, the GAMLSS method is employed to model and fit the time-varying characteristics of the parameters both in marginal and joint distributions of annual maximum streamflow series. Then, the Copula function is employed to conduct the point estimations of non-stationary design floods. Finally, the BUP technique is employed to make the interval estimations of design floods based on the point estimations obtained from the Copula function. To verify the applicability of the proposed approach in probabilistic interval estimations of design floods, this study utilizes annual maximum streamflow series of two hydrological monitoring stations in the Hanjiang River of China as a case study.

2. STUDY AREA AND MATERIALS

2.1. Study area

The Hanjiang River catchment with an area of 159,000 km^2 is located in central southern China (Figure 1). It has 1,567 km river length and 49.3 billion m^3 average annual runoff. The Hanjiang River catchment has a subtropical monsoon climate and spatial heterogeneity of water resources. Most of precipitation (70–80%) occurs in the flood season (May–September), and the annual value varies from 700 m to 1,100 mm. In the past decades, several reservoirs have been constructed in the river basin. The characteristics of 10 cascade reservoirs are listed in Table 1. The operation purposes of the 10 reservoirs include flood control, water supply (or water transfer) and hydro-power generation. The total flood control storage and regulatory storage of 10 reservoirs reach 15.4 and 21.5 billion m^3, respectively. The Danjiangkou Reservoir is the largest reservoir in the river basin and has been used as the water source of the middle route of the South-to-North Water Transfer Project since 2014 (Zhou *et al.* 2017), which has an annual water transfer demand of 9.5 billion m^3. The Danjiangkou Reservoir that is located in the middle-stream controls >58% runoff of the catchment. Besides, the Ankang monitoring station controls a catchment area of 38,600 km^2 and is located at 30 km downstream of the Ankang Reservoir, while the Huangzhuang monitoring station controls a catchment area of 142,000 km^2 and is located at 240 km downstream of the Danjiangkou Reservoir. The Shiquan Reservoir and Ankang Reservoir, which are located in the upstream of the Hanjiang River, control <7% runoff of the catchment. The reservoirs that are located in the tributaries of the Hanjiang River control >12% runoff of the catchment.

2.2. Material collection

The meteorological network monitoring daily rainfall over the Hanjiang River basin contains >100 stations in 2019, about 38 of which also gauge hourly rainfall. Significant variation of the number of monitoring stations appeared over the years, leading to relatively few long precipitation observations for a long-term (>60 years) variability detection. Consequently, 12 meteorological monitoring stations are used in this study by considering the data availability and representativeness: (1) the length of the rainfall dataset at each meteorological monitoring station is >60 years spanning from 1954 to 2018 and (2) the stations are evenly distributed in the Hanjiang River basin (Figure 1). To reveal the impacts of precipitation change and reservoir regulation on annual maximum streamflow series, this study collected the monitoring records of hydro-meteorological factors occurring from 1954 to 2018, including the accumulated annual precipitation series of 12 meteorological monitoring stations, and the annual maximum streamflow series and flood control capacity corresponding to two hydrological monitoring stations

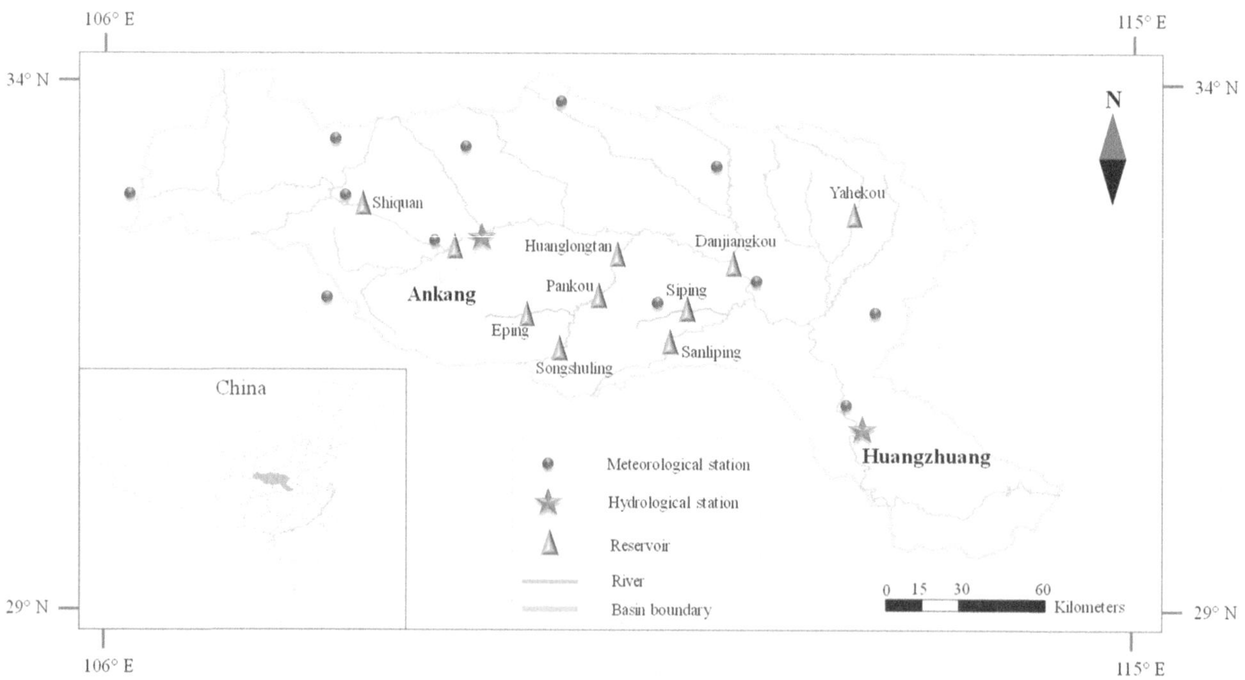

Figure 1 | Map of the Hanjiang River basin. The Ankang and Huangzhuang hydrological stations located at the upstream and downstream of the river basin, respectively.

Table 1 | Characteristics of 10 reservoirs in the Hanjiang River basin

Parameter	Reservoir									
	Shiquan	**Ankang**	**Eping**	**Songshuling**	**Pankou**	**Huanglongtan**	**Danjiangkou**	**Sanliping**	**Siping**	**Yahekou**
Main function beside flood control	Hydro-power	Hydro-power	Hydro-power	Hydro-power	Hydro-power	Hydro-power	Water supply, hydro-power	Hydro-power	Hydro-power	Hydro-power
Top of conservation pool (m)	410	330	550	394	355	247	170	416	315	180
Top of buffer pool in Summer (m)	405	325	548	394	355	247	160	403	315	175
Top of buffer pool in Autumn (m)	405	325	548	394	355	247	164	412	315	175
Total storage (million m^3)	440	2,925	302	57	2,353	1,162	2,9050	499	269	1,316
Regulatory storage (million m^3)	180	1,670	153	32	1,120	598	1,6360	211	162	1,040
Flood storage (million m^3)	180	360	153	0	0	0	1,4100	121	0	521
Installed capacity (MW)	225	850	114	50	500	510	900	70	60	13
Put into operation (year)	2002	1992	2005	2006	2011	1976	1968	2013	2006	1960

(Ankang and Huangzhuang) in the Hanjiang River basin. The annual maximum streamflow series of Ankang and Huangzhuang stations are selected from the observed daily river discharge datasets. The Ankang and Huangzhuang monitoring stations are selected as the pair of design sites in this study due to the following reasons: (1) the two stations are important for flood control in the Hanjiang River and (2) modeling the degree of dependence between two stations can adequately quantify the effects of precipitation change and reservoir regulation on basin-scale flood variability. The hydrological monitoring datasets and reservoir characteristic values related to the reservoir index (RI) can be downloaded from the Changjiang Hydrology website (http://www.cjh.com.cn/, Chinese), and the rainfall monitoring datasets can be downloaded from the China National Meteorological Information Center website (http://data.cma.cn/site/index.html, Chinese).

2.3. Pre-analysis of materials

The Kendall's τ correlation analysis was performed to determine what precipitation metrics were the most important in explaining the variability of annual maximum streamflow series. The monthly precipitation is transformed into the accumulated annual precipitation, which displays a higher Kendall's τ correlation with streamflow series as compared with the precipitation in monthly and seasonal scales. Consequently, the accumulated annual precipitation is taken as the covariate of flood non-stationarity in this study.

Figure 2 presents the time series of annual maximum streamflow, areal mean annual precipitation and flood control capacity corresponding to the two hydrological stations (Ankang and Huangzhuang) from 1954 to 2018. The Pettitt test

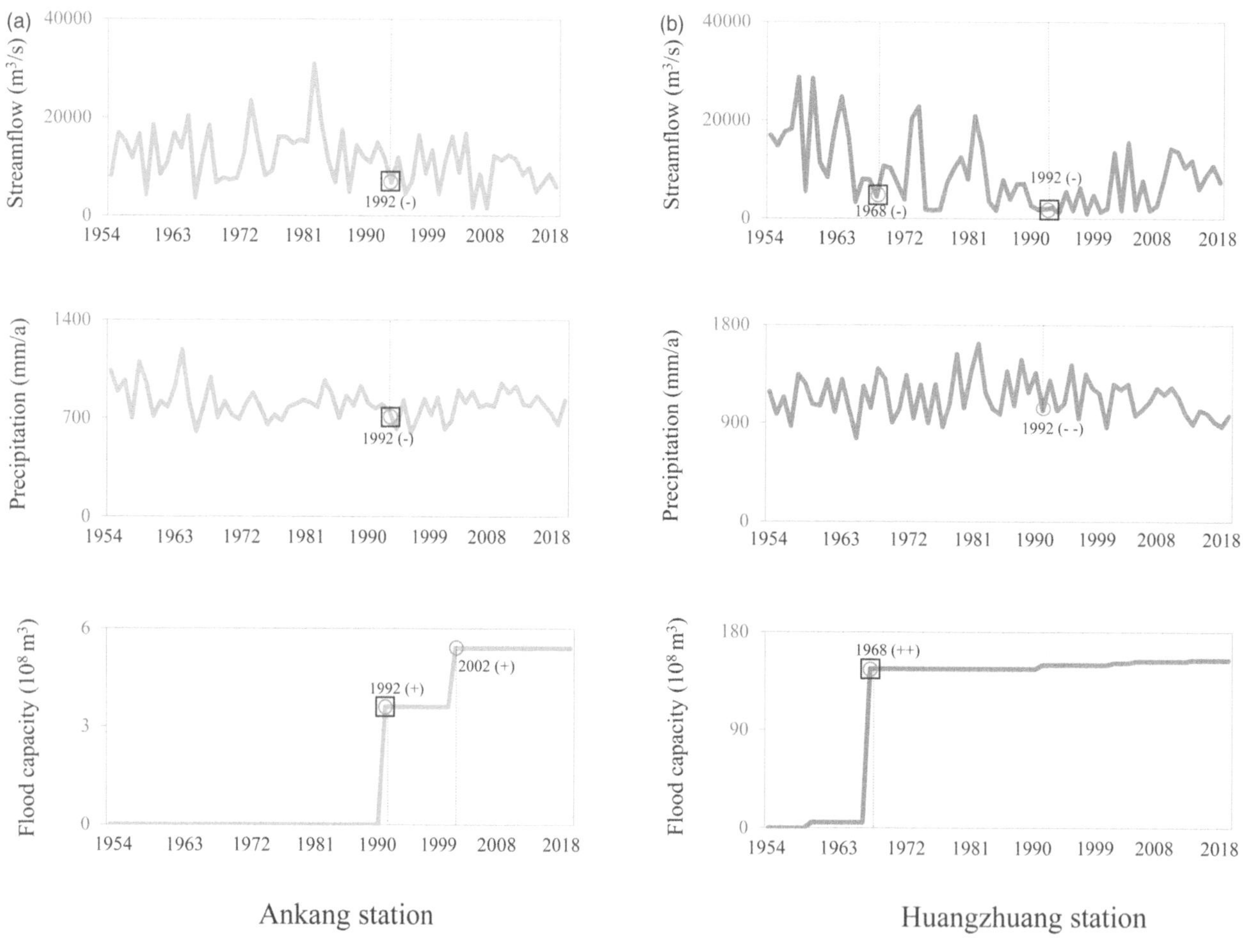

Figure 2 | Change point in the time series (mean or variance) of annual maximum streamflow, areal mean annual precipitation and flood control capacity corresponding to the Ankang and Huangzhuang hydrological stations. The precipitation is the mean accumulated annual precipitation of sub-catchment above the hydrological station. The symbol of +(++) or −(−−) indicates an increasing or decreasing change in mean or variance at 5% (10%) level, while the symbol of the circle (square) denotes mean (variance). The flood capacity of the Ankang station corresponds to the Shiquan and Ankang reservoirs, while the flood capacity of the Huangzhuang station contains 10 reservoirs (Figure 1).

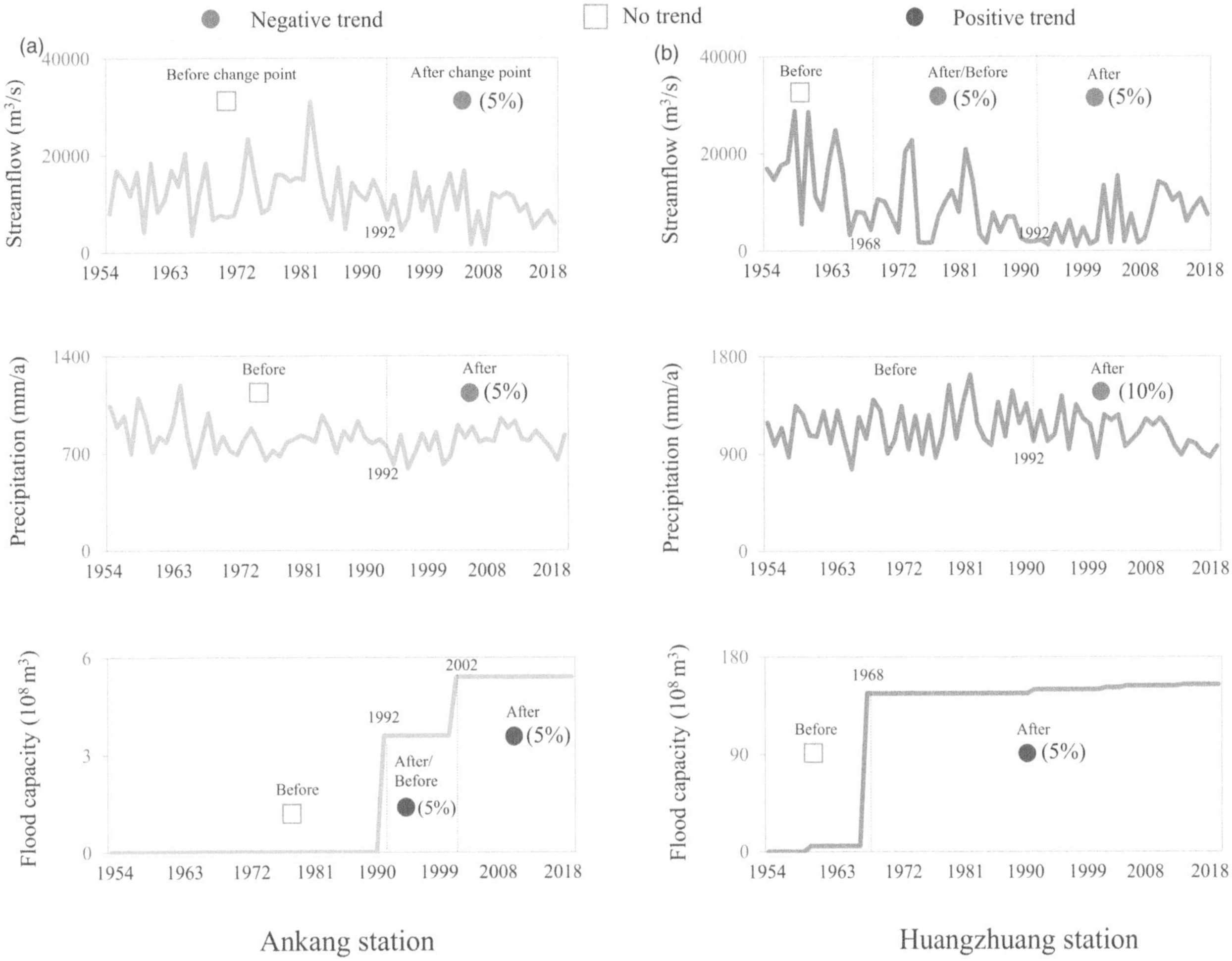

Figure 3 | Trend change in the time series of annual maximum streamflow, areal mean annual precipitation and flood control capacity corresponding to the Ankang and Huangzhuang hydrological stations. The test is significant at 5 or 10% level marked in the bracket.

(Villarini *et al.* 2009) and the Lombard Mood test (James & Matteson 2015) were used to detect change points corresponding to mean and variance, while the Mann–Kendall test (Villarini & Smith 2010) was used to detect trend change of these time series. In this study, we conducted the change point test for both mean and variance. The information from the change point test was considered to separate the time series into two subseries (before and after change point) on which the trend change analysis would be then implemented separately.

The results of Figure 2 reveal that both stations exhibit one or two change points in the mean or variance of all time series at 5 or 10% significance level. Figure 3 shows the trend change in the time series of annual maximum streamflow, areal mean annual precipitation and flood control capacity corresponding to the Ankang and Huangzhuang hydrological stations. The flood and precipitation series in two stations manifest a statistically significant decreasing trend at 5 or 10% level. Therefore, it is essential to conduct non-stationary flood frequency analysis to further quantify the impacts of precipitation change and reservoir regulation on design flood values in the Hanjiang River basin.

3. METHODS

The architecture of the integrated approach is illustrated in Figure 4, where Figure 4(a) presents the GAMLSS method, Figure 4(b) presents the Copula function and Figure 4(c) presents the BUP technique, respectively. The time-varying characteristics both in marginal and joint distribution parameters are modeled and fitted by the GAMLSS method. The point estimations of non-stationary design floods are conducted using the Copula function. The interval estimations of design floods are produced by the BUP post-processing technique. The methods used in this study are briefly introduced as follows.

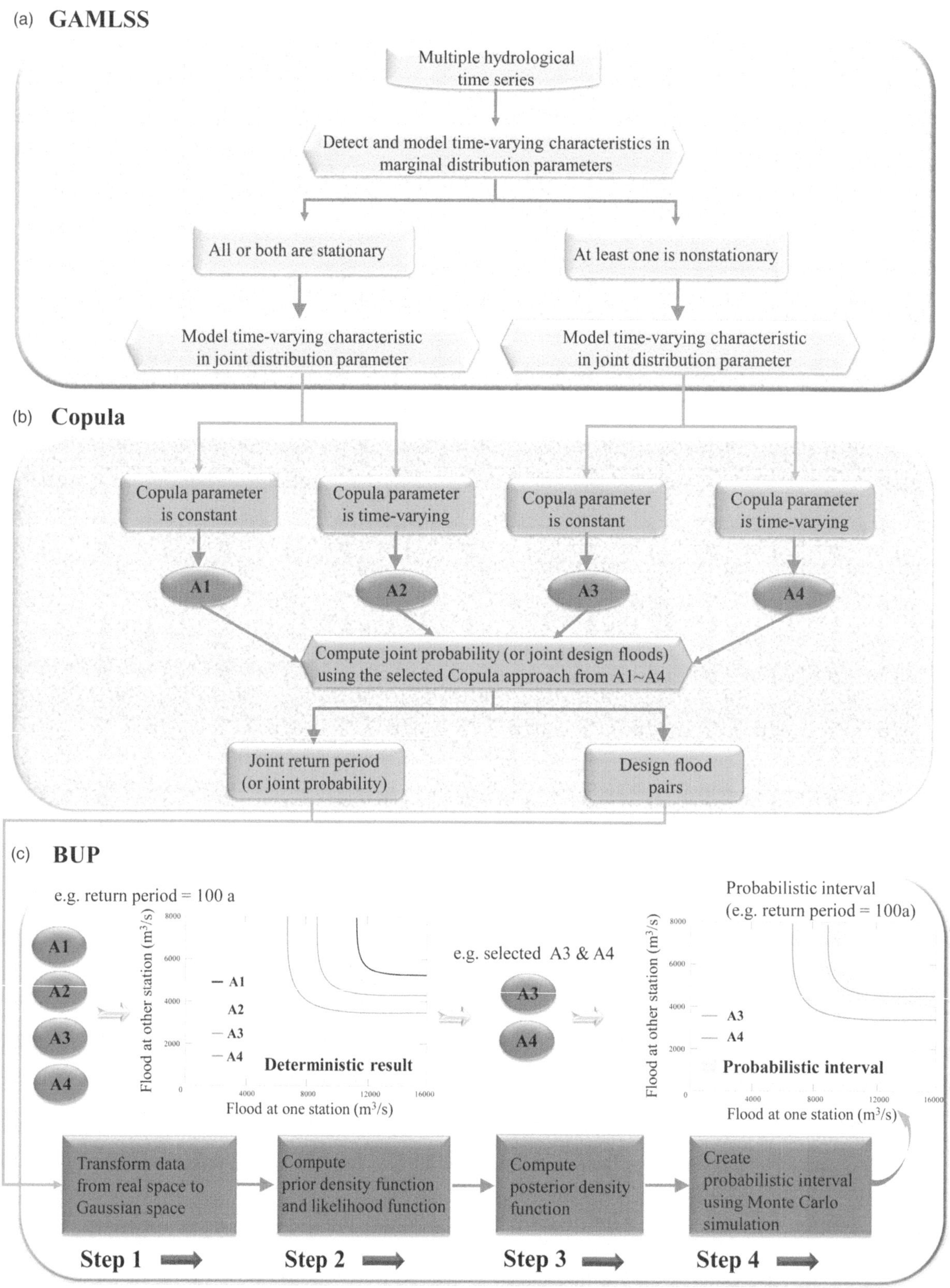

Figure 4 | Architecture of the integrated approach. (a) GAMLSS method for modeling time-varying characteristics of distribution parameters. (b) Copula function for performing point estimations of non-stationary design floods. (c) BUP technique for creating interval estimations of design floods.

3.1. GAMLSS method

The GAMLSS method is adopted to model a parametric distribution of annual maximum streamflow series with time-varying characteristics by considering the distribution parameter as a function of explanatory variables x_i^t ($i = 1, 2, \ldots, m$) with time t.

The time-varying characteristics in the two marginal distributions can be classified into two kinds, i.e. both are stationary and at least one is non-stationary (Figure 4(a). Because the response variable relaxes a distribution followed the exponential family in the GAMLSS method, this method allows for a general distribution function even highly skewed. More commonly when using the GAMLSS method, two-parameter distributions are used to develop non-stationary models. The two-parameter distributions are less complicated and if the parameters of the distribution are best modeled by physical covariates, it is less likely that higher-order distributions are needed to explain the variability of annual maximum streamflow series (Villarini *et al.* 2009, 2012; Jiang *et al.* 2015; Xiong *et al.* 2015b). Therefore, this study assumes time-varying characteristics resided in the location and scale parameters, whereas the shape parameter is assumed to be constant with time in all distributions.

$$g_1(\alpha_i^1) = \alpha_{10} + \sum_{i=1}^{m} \alpha_{1i} x_i^t \tag{1}$$

$$g_2(\alpha_i^2) = \alpha_{20} + \sum_{i=1}^{m} \alpha_{2i} x_i^t \tag{2}$$

where x_i^t is the i-th ($i = 0, 1, 2, \ldots, m$) explanatory variable at t time. m is the number of covariates (i.e. explanatory variables). $g(\cdot)$ is the log link function recognizing that the annual maximum streamflow series may be skewed. $\alpha = (\alpha_i^1, \alpha_i^2)$ ($i = 0, 1, 2, \ldots, m$) is the vector of distribution parameters accounting for location and scale, where $\alpha_i^1 = [\alpha_{10}, \alpha_{1i}]^T$ and $\alpha_i^2 = [\alpha_{20}, \alpha_{2i}]^T$.

This study concentrates on modeling the impacts of climate variability and reservoir regulation on flood non-stationarity in a river basin. Considering the precipitation as the proxy for flood-generating mechanism as well as the RI as the proxy for anthropogenic regulation (Jiang *et al.* 2015; Agilan & Umamahesh 2017), these drivers are taken as the explanatory variables of the marginal distribution parameters in this study. The RI is described as follows:

$$\mathrm{RI} = \sum_{i=1}^{N} \left(\frac{A_i}{A_T}\right) \cdot \left(\frac{V_i}{V_T}\right) \tag{3}$$

where A_i and A_T are the drainage area regulated by the ith reservoir and the total drainage area monitored by the river streamflow gauge station, respectively. V_i and V_T are the flood capacity of the ith reservoir and the total reservoir flood capacity in the river streamflow gauge station, respectively. N is the number of reservoirs.

For comparative purpose, we considered several three-parameter distributions and two-parameter distributions to model the distribution of annual maximum streamflow series in this study (Table 2).

All the aforementioned computations are conducted in R (https://www.r-project.org/) using the freely available GAMLSS package (Stasinopoulos *et al.* 2008).

3.2. Copula functions for point estimations of design floods

After modeling the time-varying characteristics of two marginal distributions, the Copula functions are used to compute the joint probability (or joint design floods) under the four scenarios of the parameters in the marginal and joint distributions (Figure 4(b).

Three widely used Archimedean Copula functions (Table 3) are employed as the candidates for fitting joint probability distributions (Xiong *et al.* 2014; Yu *et al.* 2014; Li *et al.* 2019). The time-varying parameter (θ_c) of the Copula function can be modeled as a function of covariates x_i^t ($i = 1, 2, \ldots, m$) and described as follows:

$$g_c(\theta_c^t) = \beta_0 + \sum_{i=1}^{m} \beta_i x_i^t \tag{4}$$

$$C(u_1^t, u_2^t | \theta_c^t) = C[F_1(y_1^t | \theta_1^t), F_2(y_2^t | \theta_2^t) | \theta_c^t] \tag{5}$$

Table 2 | Univariate distribution functions for fitting marginal distributions

Distribution	Probability distribution function (pdf)	Range	Parameter
Gaussian	$f(x) = \frac{1}{\sqrt{2\pi}\sigma}\exp\left[-\frac{(x-\mu)^2}{2\sigma^2}\right]$	$-\infty < x < +\infty$	μ, σ
Gamma	$f(x) = \frac{\beta^\alpha}{\Gamma(\alpha)}x^{\alpha-1}\exp(-\beta x)$	$x > 0$	α, β
Gumbel	$f(x) = \alpha\exp[-\alpha(x-\mu) - e^{-\alpha(x-\mu)}]$	$-\infty < x < +\infty$	α, μ
GEV	$f(x) = \frac{1}{\sigma}\left[1+\nu\left(\frac{x-\mu}{\sigma}\right)\right]^{(-1/\nu)-1}\cdot\exp\left\{-\left[1+\nu\left(\frac{x-\mu}{\sigma}\right)\right]^{-1/\nu}\right\}$	$-\infty < x < +\infty$	μ, σ, ν
Person type III	$f(x) = \frac{\beta^\alpha}{\Gamma(\alpha)}(x-\mu)^{\alpha-1}\exp[-\beta(x-\mu)]$	$x > \mu$	α, β, μ
Log-Weibull	$f(x) = \frac{1}{\alpha(x-\mu+1)}\left[\frac{\ln(x-\mu+1)}{\alpha}\right]^{\beta-1}\cdot\exp\left\{-\left[\frac{\ln(x-\mu+1)}{\alpha}\right]^{\beta}\right\}$	$x > \mu$	α, β, μ

Table 3 | Bivariate Archimedean Copula functions for fitting joint distribution

Copula function	Joint distribution function	Parameter
Gumbel–Hougaard (GH)	$C(u_1, u_2\|\theta) = \exp\{-[(-\ln u_1)^\theta + (-\ln u_2)^\theta]^{1/\theta}\}$	[a] $\tau = 1 - \frac{1}{\theta}$ $\theta \geq 1$
Frank	$C(u_1, u_2\|\theta) = -\frac{1}{\theta}\ln\left[1 + \frac{[\exp(-\theta u_1) - 1][\exp(-\theta u_2) - 1]}{\exp(-\theta) - 1}\right]$	$\tau = 1 + \frac{4}{\theta}\left[\frac{1}{\theta}\int_0^\theta \frac{t}{\exp(t)}d_t - 1\right]$ $-\infty < \theta < +\infty$
Clayton	$C(u_1, u_2\|\theta) = (u_1^{-\theta} + u_2^{-\theta} - 1)^{-1/\theta}$	$\tau = \frac{\theta}{2+\theta}$ $\theta > 0$

[a] τ is the Kendall's coefficient.

where $g_c(\cdot)$ and $C(\cdot)$ are the link function and the Copula function, respectively. $[\beta_i]^T$ (i=0, 1, 2, …, m) is the parameter vector of covariates. The link function $g_c(\cdot)$ relies on the value of parameters in Copula functions, i.e. for Frank Copula function ($\theta_c \in R$), $g_c(\theta_c) = \theta_c$, whereas for Gumbel–Hougaard and Clayton Copula functions ($\theta_c > 0$), $g_c(\theta_c) = \ln(\theta_c)$. u_1^t and u_2^t are the two marginal probabilities of the Copula function at t time, respectively, which should be both uniformly distributed on [0, 1]. $F(\cdot)$ is the cumulative distribution function with non-exceedance probability. y_1^t and y_2^t are the annual maximum streamflow series of two marginal distributions at t time, respectively. θ_c^t is the time-varying parameter of the Copula function at t time. θ_1^t and θ_2^t are the time-varying parameters of two marginal distributions at t time, respectively.

In this study, the Inference Function for Margins method (Joe 1997) is used to estimate the parameters of the time-varying joint distribution. In addition, the evaluation of marginal distributions and joint distribution is conducted by comparing the values of Kolmogorov–Smirnov (K–S) test, Akaike Information Criterion (AIC, Akaike 1974) and diagnostic plots (e.g. worm plots). Owing to the higher complexity of a joint distribution (e.g. Copula functions), the selection of a particular joint distribution can be further examined by computing the first four moments of the residuals and their Filliben correlation coefficient (Villarini *et al.* 2012). The particular distribution with the minimum K–S indicator, the minimum AIC value and rather flat worm plots, as well as the statistical properties of the residuals close to the standard normal distribution, is selected. For more details about distribution selection and fitting, the interested reader is pointed to the references (Stasinopoulos *et al.* 2008; Villarini *et al.* 2012).

And then, the time-varying Copula function with time-varying marginal distributions is employed to calculate joint probability and the return period of two annual maximum streamflow series (i.e. flood pairs). The joint probability and the return period of flood pairs are defined as follows:

$$P_{max} = P(Y_1 \geq y_1 \cap Y_2 \geq y_2) = 1 - F_1(y_1|\theta_1^t) - F_2(y_2|\theta_2^t) + C[F_1(y_1|\theta_1^t), F_2(y_2|\theta_2^t)|\theta_c^t] \tag{6a}$$

$$T = \frac{1}{1 - P(Y_1 \geq y_1 \cap Y_2 \geq y_2)} \tag{6b}$$

where P_{max} and T are the joint probability and the return period corresponding to flood pairs. $P(\cdot)$ is the joint probability function. Y_1 and Y_2 are the two marginal distributions, respectively. y_1 and y_2 are the observations of two annual maximum streamflow series, respectively.

3.3. BUP for the probabilistic interval estimation of design floods

In this study, the BUP proposed by Krzysztofowicz (1999) is applied to model the non-linear dependence between observed and estimated data (i.e. design flood values) at each joint probability p (e.g. p starts from 0.01 up to 99.99% with time step 0.01%) step by step. The implementation procedures of the BUP (Figure 4(c)) for estimating the probabilistic interval of design flood values consist of the following four steps (Krzysztofowicz 1999):

- *Step 1*: Implement data transformation for converting observed (H_p) and estimated (S_p) flood pairs from the real space to the Gaussian space.
- *Step 2*: Compute the prior density and likelihood functions of flood pairs.
- *Step 3*: Calculate the posterior density function of flood pairs.
- *Step 4*: Conduct data transformation for converting datasets from the Gaussian space to the real space and then execute the Monte Carlo simulation to produce the probabilistic interval estimation of design flood values.

4. RESULTS AND DISCUSSION

The results and findings were presented in the order of the time-varying characteristics in distribution parameters (Section 4.1), flood frequency analysis using the Copula function (Section 4.2) and probabilistic interval estimations of design floods using the BUP technique and summarization (Section 4.3), and were detailed as follows.

4.1. Time-varying characteristics in distribution parameters

Table 4 summarizes the results of the two time-varying marginal distributions with the precipitation and the RI as the covariates corresponding to the flood-generating mechanism and reservoir regulation in the distribution parameters based on the GAMLSS method. The results point out that the Gumbel (GU) distribution and generalized extreme value (GEV) distribution can be considered as the best-fitted distributions for Q_a and Q_h, respectively, owing to the minimum values of K–S and AIC indicators. We also tested each covariate in the case separately, in which these experiment schemes created larger values of K–S and AIC indicators as compared with the results of Table 4. Therefore, when time-varying marginal distribution considers precipitation and RI simultaneously as covariates, the GU and GEV distributions have a good quality in fitting the two annual maximum streamflow series (Q_a and Q_h), respectively.

Table 4 | Results of the two time-varying marginal distributions with the precipitation and the RI as the covariates of flood-generating mechanism and reservoir regulation in the distribution parameters under non-stationary assumption

Series	Distribution	Distribution parameters			K–S statistic indicator[a]	AIC value
		μ	σ	ν		
Q_a	Gaussian	11,595	$\exp(-721+1.08P_a-14.59RI_a)$	–	0.132	459
	Gamma	11,418	$\exp(-522+0.94P_a-10.33RI_a)$	–	0.134	448
	Gumbel	10,107	$\exp(-339+1.11P_a-15.23RI_a)$	–	0.111	355
	GEV	9,817	$\exp(-283+0.76P_a-16.05RI_a)$	0.22	0.130	460
	Person type III	11,507	$\exp(-610+1.25P_a-16.58RI_a)$	0.35	0.146	502
	Log-Weibull	183	$\exp(-138+2.09P_a-10.24RI_a)$	0.30	0.195	531
Q_h	Gaussian	$\exp(-303+0.75P_h-10.01RI_h)$	0.35	–	0.148	743
	Gamma	$\exp(-219+1.02P_h-8.40RI_h)$	0.33	–	0.153	807
	Gumbel	$\exp(-412+0.91P_h-9.05RI_h)$	$\exp(-650+1.19P_a-16.61RI_a)$	–	0.131	726
	GEV	$\exp(-406+1.09P_h-15.84RI_h)$	0.38	0.27	0.115	618
	Person type III	$\exp(-452+0.78P_h-11.10RI_h)$	0.32	0.48	0.144	765
	Log-Weibull	$\exp(-229+0.84P_h-14.23RI_h)$	3.11	0.35	0.173	794

[a]The K–S test is performed at a significance level of 0.05. The null hypothesis states that the empirical distribution fits the theoretical distribution. If the value of the K–S test indicator is smaller than the value of $D(65, 0.05)$ ($= 1.36/\sqrt{65} = 0.168$), the null hypothesis would not be rejected. The variables P_a, P_h, RI_a and RI_h are the precipitation and the RI corresponding to the Ankang and Huangzhuang stations, respectively.

The change point and trend change tests (Figures 2 and 3) also demonstrate that the physical reason for non-stationary streamflow series would be attributed to precipitation change and reservoir regulation. From the perspective of flood-generating mechanism, near-record flood peaks are closely related to the subtropical monsoon climate, where such weather feature dominates the variability of precipitation in the Hanjiang River. From the perspective of reservoir regulation, Shiquan and Ankang reservoirs control 6.8% runoff of the catchment, while the rest reservoirs control >70% runoff.

In Figure 5, if all points fall within the 95% confidence interval (between two dashed semicircles), the marginal distribution has a good fitting quality. The results of worm plots (Figure 5) and the K–S statistic test (Table 4) indicate that the selected marginal distributions have a good fitting quality. Therefore, the factors of precipitation and RI can be selected as the covariates to fit the distributions of annual maximum streamflow series of Ankang and Huangzhuang stations.

Table 5 presents the results of three Copula functions with constant parameter (A3) and time-varying parameter (A4) for modeling the dependence between two streamflow series (Q_a and Q_h) at Ankang and Huangzhuang stations. The results reveal that under the time-varying dependence (scenario A4), the GH Copula function has the smallest values of K–S and AIC indicators, while the values of K–S and AIC indicators under the scenario A4 are smaller than those of the scenario A3. The comparison between two scenarios (A3 and A4) demonstrates that the GH Copula function with the climate and

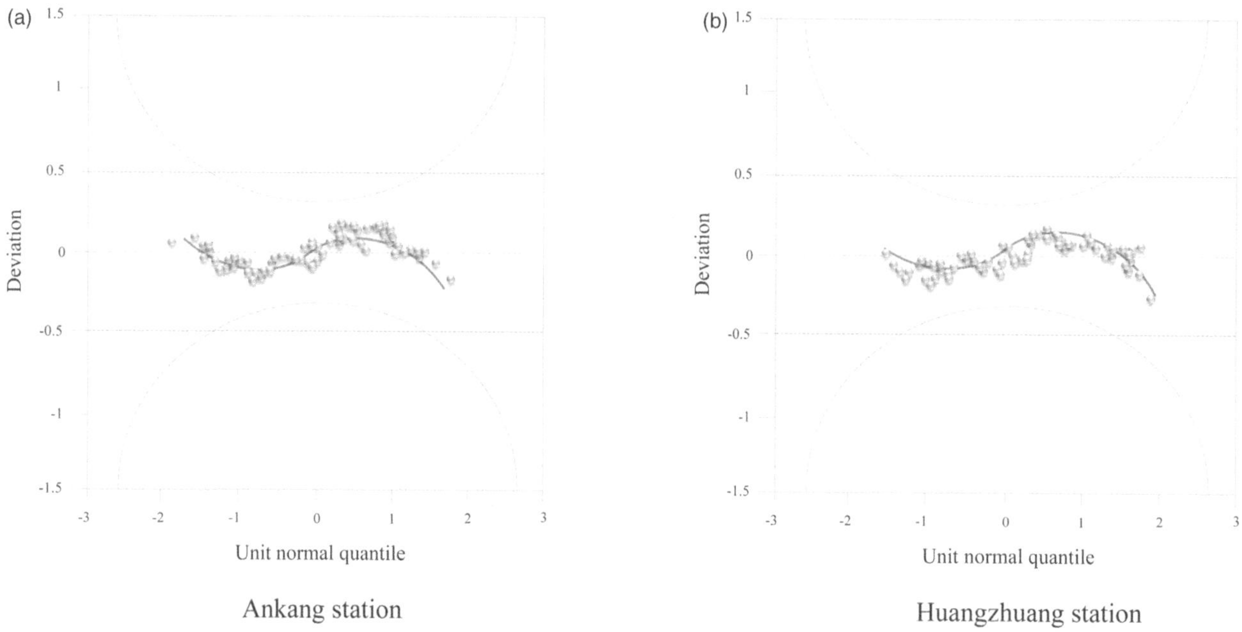

Figure 5 | Worm plots in the goodness-of-fit test for the two time-varying marginal distributions with the precipitation and RI as the climate and anthropogenic covariates of the distribution parameters.

Table 5 | Results of three Copula functions with the constant (A3) parameters and the time-varying (A4) parameters for modeling the dependence between two annual maximum streamflow series

Scenario	Copula	Parameter θ_c	K–S statistic indicator[a]		AIC value	
			Variable X_1	Variable X_2	Variable X_1	Variable X_2
A3	GH	2.70	0.127	0.131	43	55
	Frank	7.85	0.144	0.157	57	67
	Clayton	3.46	0.172	0.175	71	80
A4	GH	$\exp(-414+0.56P_h-0.73RI_h)$	0.119	0.122	33	39
	Frank	$\exp(-320+0.18P_h-1.45RI_h)$	0.149	0.154	35	49
	Clayton	$\exp(-257+0.47P_h-0.69RI_h)$	0.173	0.171	53	67

[a]The K–S test is performed at a significance level of 0.05. The null hypothesis states that the empirical distribution fits the theoretical distribution. If the value of the K–S test indicator is smaller than the value of $D(65, 0.05)$ ($=1.36/\sqrt{65}=0.168$), the null hypothesis would not be rejected. The variables X_1 and X_2 are the probability integral transformations for streamflow series Q_a and Q_h at Ankang and Huangzhuang stations, respectively.

anthropogenic covariates can effectively model the non-stationarity in parameters of the joint distribution of two streamflow series.

The first four moments of the residuals (i.e. mean, variance, skewness and kurtosis) and their Filliben correlation coefficients (Table 6) and the worm plots (Figure 6) further point out that the selected GH Copula function with the time-varying parameter (scenario A4) has a satisfactory fitting quality to model the dependence structure between Q_a and Q_h.

Table 6 | Summarization of the first four moments of the residuals and their Filliben correlation coefficients for modeling the dependence between two annual maximum streamflow series

		Variable X_1				
Scenario	**Copula**	**Mean**	**Variance**	**Skewness**	**Kurtosis**	**Filliben**
A3	GH	0.01	1.02	0.05	2.11	0.985
	Frank	0.02	1.03	0.08	2.34	0.977
	Clayton	−0.02	1.05	0.11	2.68	0.968
A4	GH	0.01	1.03	0.01	2.08	0.995
	Frank	0.01	1.02	0.05	2.17	0.980
	Clayton	−0.02	1.07	0.05	2.52	0.971
		Variable X_2				
Scenario	**Copula**	**Mean**	**Variance**	**Skewness**	**Kurtosis**	**Filliben**
A3	GH	−0.01	1.02	0.03	2.18	0.976
	Frank	−0.02	1.03	0.05	2.40	0.971
	Clayton	0.02	1.04	0.08	2.75	0.962
A4	GH	0.00	1.00	0.01	2.12	0.989
	Frank	−0.01	1.02	0.03	2.23	0.979
	Clayton	0.03	1.05	0.07	2.58	0.971

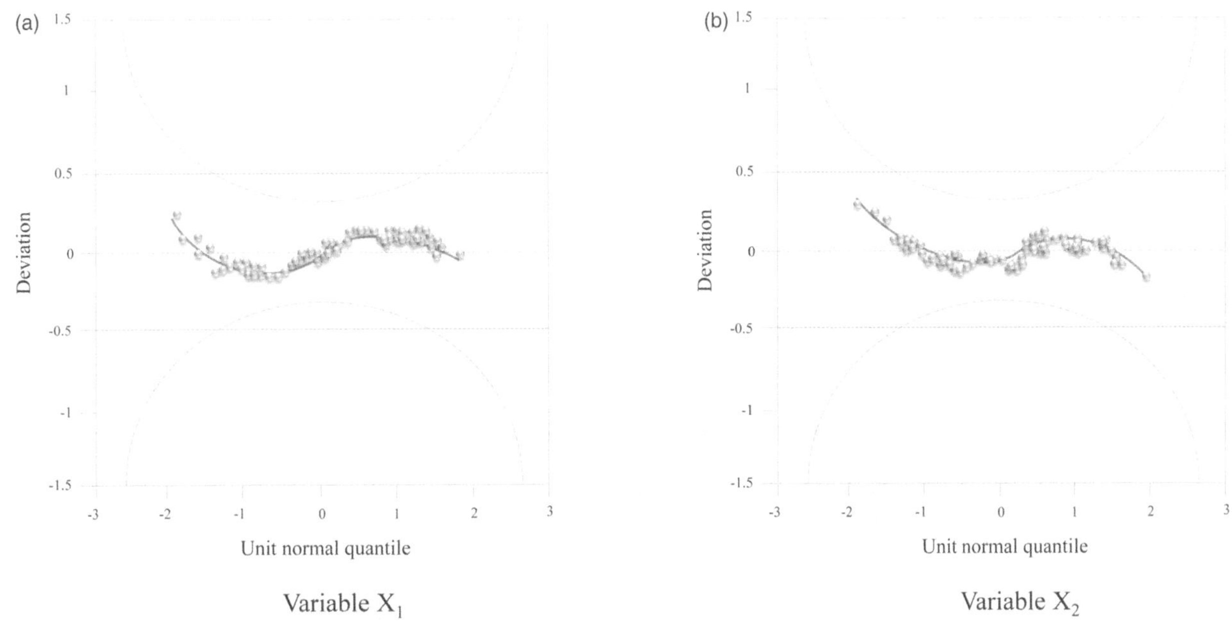

Figure 6 | Worm plots in the goodness-of-fit test for the GH Copula function with the precipitation and the RI as the climate and anthropogenic covariates of the time-varying dependence parameter (scenario A4). If all points fall within the 95% confidence interval, the two probability integral transformations (X_1 and X_2) are both uniformly distributed on [0, 1].

4.2. Point estimation of design floods using the Copula function

After determining the marginal distributions and the Copula function, the time-varying joint distribution of two annual maximum streamflow series (Q_a and Q_h) can be formulated as follows:

$$C_{Y_1,Y_2}(Q_a^t, Q_h^t) = C_{GH}[F_{GU}(Q_a^t|\boldsymbol{\theta}_a^t), F_{GEV}(Q_h^t|\boldsymbol{\theta}_h^t)|\theta_c^t] \tag{7}$$

where $F_{GU}(\cdot)$ and $F_{GEV}(\cdot)$ are the GU and GEV distributions of two streamflow series (Q_a and Q_h), respectively. $C_{GH}(\cdot)$ is the GH Copula function. The two marginal distribution parameters are expressed as follows:

$$\boldsymbol{\theta}_a^t = (10{,}107, \exp(-339 + 1.11P_a^t - 15.23\mathrm{RI}_a^t))^T \tag{8}$$

$$\boldsymbol{\theta}_h^t = (\exp(-406 + 1.09P_a^t - 15.84\mathrm{RI}_a^t),\ 0.38,\ 0.27)^T \tag{9}$$

The dependence parameter of θ_c^t of the GH Copula function is expressed as follows:

$$\theta_c^t = \exp(-414 + 0.56P_h^t - 0.73\mathrm{RI}_h^t) \tag{10}$$

Figure 7 shows the results of flood frequency analysis under the two scenarios (A3 and A4). It is interesting to find that the difference in joint return periods of two scenarios is significant ($T_{A3} < T_{A4}$) after 1967, especially for medium–high flood magnitudes. Being incorporated with climate and anthropogenic covariates, the time-varying property of the Copula function not only can model the dependence between variables, but also mitigate the under-estimation of joint return periods. Despite the fact that flood non-stationarity of the Hanjiang River can be attributed to both climate change and reservoir regulation, the

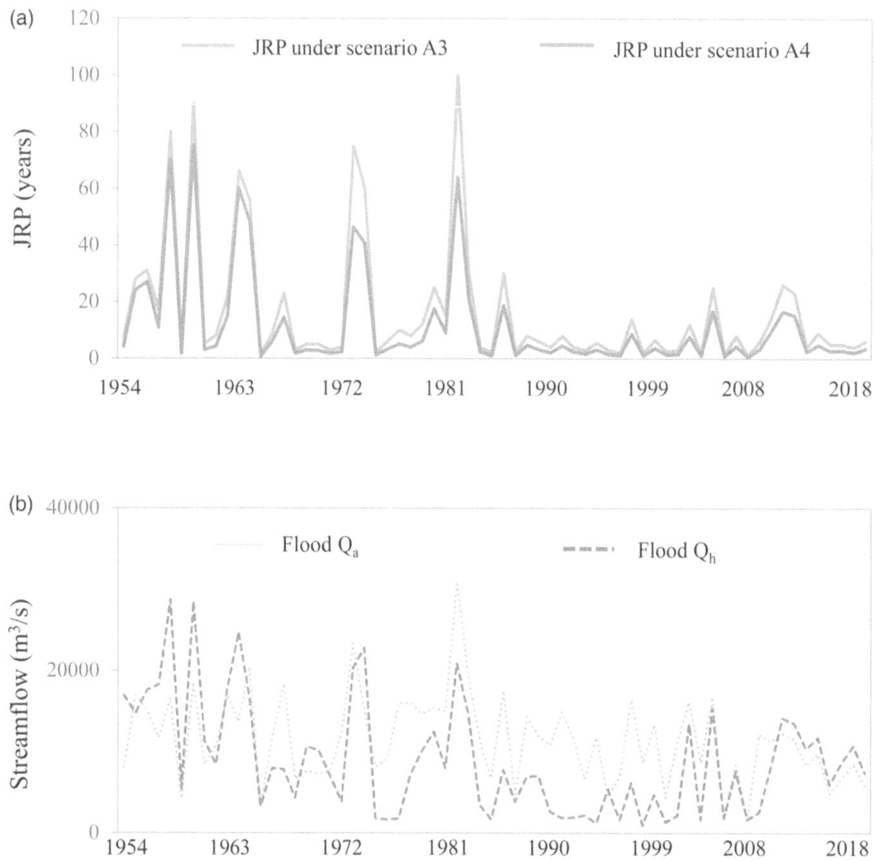

Figure 7 | Joint return period (JRP) and observed annual maximum streamflow series. (a) JRP of observed flood pair (Q_a and Q_h) under the two scenarios (A3 and A4). (b) Observed streamflow series at the Ankang and Huangzhuang stations.

significant variation after 1967 is mainly due to the regulation of the largest reservoir (i.e. Danjiangkou Reservoir). The Danjiangkou Reservoir is able to control >58% runoff of the Hanjiang River catchment, since it was put into operation in 1967. Considering climate change and reservoir regulation, the time-varying characteristics in marginal distributions and Copula function have significant impacts on joint return periods of flood pairs (Q_a and Q_h).

Consider that the parameters in Equations (8)–(10) have time-varying characteristics, 1,000 simulations of design flood pairs corresponding to a given joint return period under the scenario A4 were performed to further investigate the impacts of precipitation change and reservoir regulation on the joint return period of annual maximum streamflow series (Q_a and Q_h). The median values of design flood pairs corresponding to a given joint return period (100 years) were computed in accordance with the joint distribution of each time segment.

Figure 8 displays the isolines of the design flood pairs corresponding to the joint return period (100 years) at three-time segments under the scenarios A3 and A4. The results indicate that (1) from the first time segment (1954–1967) to the second time segment (1968–1991), the isolines of the joint return period under two scenarios (A3 and A4) move to the left, owing to the decreasing mean value of annual maximum streamflow series Q_h (Huangzhuang station); (2) and then, from the second time segment (1968–1991) to the third time segment (1992–2018), the isolines of the joint return period under two scenarios (A3 and A4) drastically move downward, owing to the decreasing coefficient of variation in both annual maximum streamflow series Q_a (Ankang station) and Q_h (Huangzhuang station). The coefficient of variation denotes the extent of variability in relation to the mean of the population. For the scenario A3, it is easy for the Copula function with a constant parameter to over-estimate the coefficient of variation without modeling the time-varying characteristics. For the scenario A4, both marginal and joint distributions have been modeled with the time-varying parameters to adequately quantify the coefficient of variation. The decreasing coefficient of variation in two streamflow series can be in a large measure attributed to the impact of climate change and reservoir regulation.

There are both lateral translations (reduction in flow magnitudes) and vertical decreases in flow magnitudes. The reasons for inducing both lateral and vertical translations consist of (1) the largest Danjiangkou Reservoir in the Hanjiang river has been put into operation since 1967 and largely reduces the mean value of annual maximum streamflow series Q_h

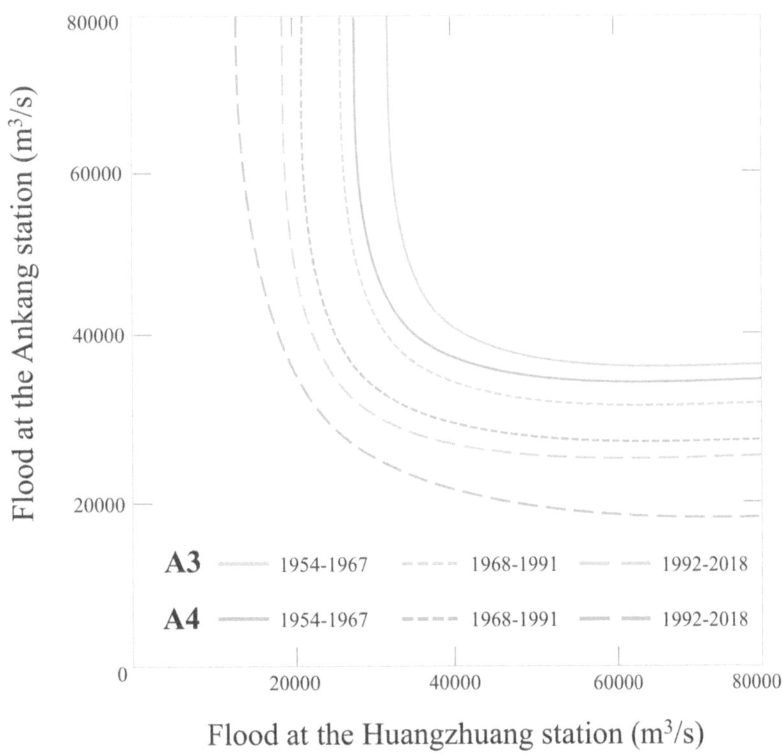

Figure 8 | Isolines of the design flood pairs with the given joint return period (100 years) corresponding to three-time segments of 1954–1967 (14 years), 1968–1991 (24 years) and 1992–2018 (27 years) under the scenarios A3 and A4.

(Huangzhuang station) and (2) due to the completion of the rest of nine reservoirs during 1991–2018 and the decreasing trend of precipitation after 1991, the variances of streamflow series both in two stations decrease significantly. The changing covariates of climate and reservoir regulation (Figures 3 and 4) would adequately explain the physical meaning of the combination of lateral and vertical translations.

4.3. Probabilistic interval estimations of design floods using the BUP technique

We further assessed the impacts of time-varying Copula function on the uncertainty of the joint probability between two annual maximum streamflow series (Q_a and Q_h) based on Quantile–Quantile (QQ) plots. Figure 9 presents the QQ plots of probabilistic flood frequency analysis corresponding to three-time segments (1954–1967, 1968–1991 and 1992–2018) using the BUP approach under two scenarios (A3 and A4).

It can be seen from Figure 9 that (1) from the first time segment (1954–1967) to the second time segment (1968–1991) and the third time segment (1992–2018), the differences between the two QQ plotlines under two scenarios (A3 and A4) become more and more significant and (2) for all three-time segments, the QQ plots corresponding to scenario A4 are closer to the theoretical 1:1 line, in comparison to those of the scenario A3. That is to say, the non-stationary Copula function (scenario A4) produces the smaller biases of the joint probability than the stationary Copula function (scenario A3). The results demonstrate that the BUP approach can effectively quantify the uncertainty of flood frequency analysis owing to its better agreement to the theoretical 1:1 line.

Figure 10 presents the probabilistic interval estimations of design floods for the given joint return period (100 years) corresponding to the three-time segments (1954–1967, 1968–1991 and 1992–2018) using the BUP approach under the two scenarios (A3 and A4). The results reveal that (1) for the first time segment (1954–1967), the probabilistic interval of design floods created by the BUP approach under the scenario A4 (non-stationary Copula) approximates to that under the scenario A3 (stationary Copula), because the impacts of climate change and reservoir regulation on flood non-stationarity are not significant in the first time segment (Figures 3 and 4) and (2) from the first time segment (1954–1967) to the next two time segments (1968–1991 and 1992–2018), the probabilistic interval of design floods created by the BUP approach under the scenario A4 is smaller than that of the scenario A3, because the impacts of climate change and reservoir regulation on flood non-stationarity are becoming more and more significant in the next two time segments (Figures 3 and 4). The results demonstrate that the BUP post-processing technique can effectively quantify the uncertainty of probabilistic interval estimation of non-stationary design floods by decreasing the probabilistic distribution to a small range.

5. DISCUSSION

Because the one-on-one relationship between the return period and the return level that exists in the univariate case does not exist in the non-stationary and the multivariate case, it is important to provide an indication on how design values (i.e. flood pairs) can be inferred in such a high-dimensional problem. Take the lines A–C under the scenario A4 in Figure 10, for example, the probabilistic interval estimations can provide the maximal and minimal design values to increase the reliability in the planning and design of hydraulic engineering projects. Furthermore, there is no doubt that the hydro-meteorological environment is experiencing variability due to a changing world. Accordingly, the proposed methodology and the pre-experience maps (e.g. point estimation, isolines of design flood pairs and interval estimation) of hydrological uncertainty presented in this study are useful tools to improve the operation process, in which the predictive hydrological design is essential because the hydrological frequency and the return period will be closely associated with the variability of hydro-meteorological conditions.

The results of this study may be affected by the combined methods. For instance, the GAMLSS method needs to consider the uncertainty of model parameters containing various possible options and combinations with respect to explanatory variables and link functions. In addition, the curse of dimensionality cannot be avoided: Copula function fitting and parameter estimation are more demanding and computationally intensive in higher dimensions. Finally, the BUP application needs transform point estimations from the original space to the Gaussian space before producing interval estimations. Such data space transformation is bound to induce information loss. Therefore, more modular design, parameter estimation and hybridization are imminent to quantify and reduce the uncertainty encountered in the interval estimation of design floods.

The proposed methodology is scalable and transferable, with capacity to address problems ranging from non-stationary hydrological frequency analysis to uncertainty assessment for local and global regions of concerns. There are quite many physical attributions of hydro-meteorological non-stationarity. This study is only a case that utilizes the integrated approach to quantify the impacts of precipitation and reservoir regulation on flood non-stationarity. Notwithstanding these promising

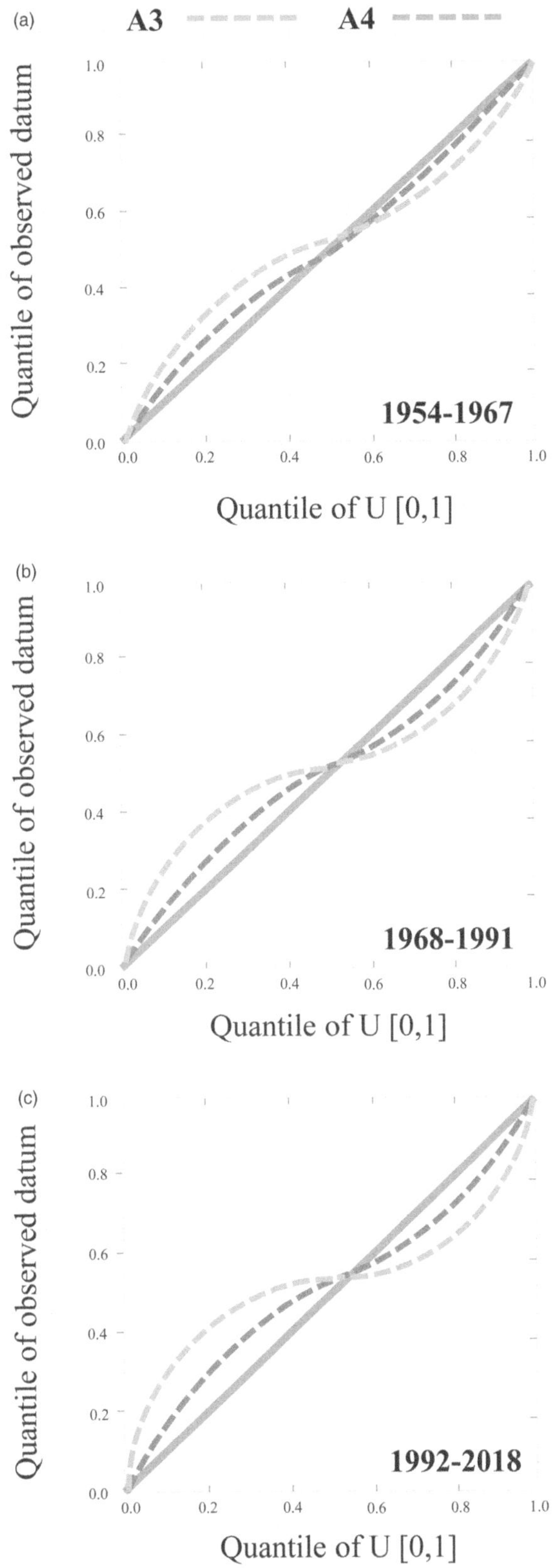

Figure 9 | QQ plots for the probabilistic estimation of the joint probability between two annual maximum streamflow series (Q_a and Q_h) using the GH Copula function corresponding to three-time segments of (a) 1954–1967, (b) 1968–1991 and (c) 1992–2018 under the scenarios A3 and A4.

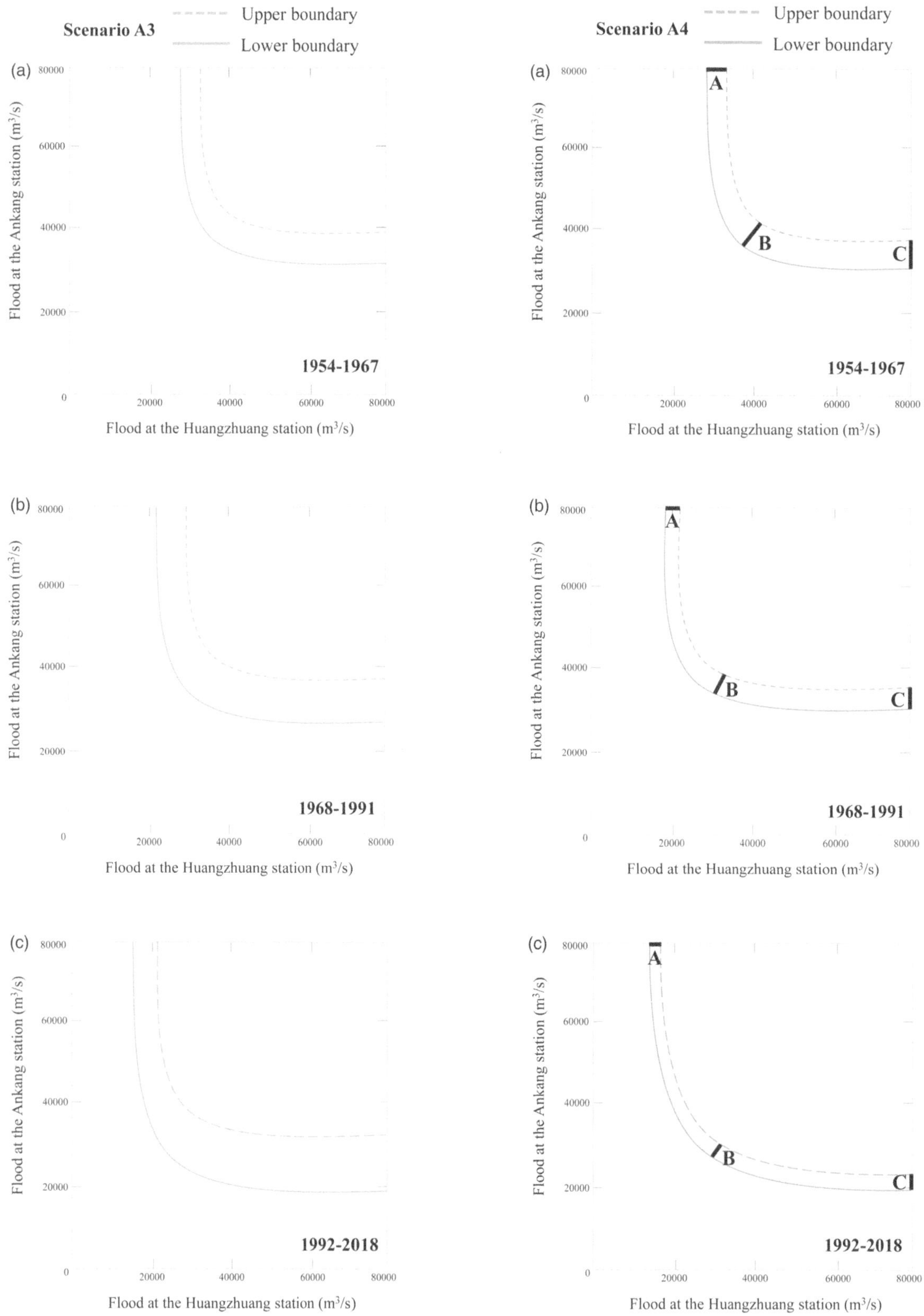

Figure 10 | Probabilistic interval estimations of design flood pairs with the given joint return period (100 years) corresponding to the three-time segments ((a) 1954–1967, (b) 1968–1991 and (c) 1992–2018) under the GH Copula function with the constant dependence parameter (scenario A3) and the time-varying dependence parameter (scenario A4). The number of simulations for estimating probabilistic interval is 1,000.

achievements, future research can be executed to investigate the far-reaching effects of climate change on future flood design values using climate model outputs.

6. CONCLUSIONS

This study proposes an integrated approach that combines the GAMLSS method, the Copula function and the BUP post-processing technique to bring about the probabilistic interval estimation of design floods. The capability of the proposed approach is verified at the Ankang and Huangzhuang hydrological stations in the Hanjiang River basin of China based on long-term (1954–2018) observed daily hydrological datasets. The results reveal that the proposed approach not only adequately model the time-varying characteristics of parameters in both marginal and joint distributions of annual maximum streamflow series, but also effectively increase the reliability of probabilistic interval estimation of design floods encountered in high climate variability and reservoir regulation conditions. The findings of this study are summarized as follows:

1. At a significance level of 0.05 or 0.1, the years 1968 and 1992 are identified as change points with significant and moderate decreasing trends (mean or variance), respectively. The annual maximum streamflow series of the Ankang station have significantly changed in 1992, due to the regulation of the Ankang Reservoir meanwhile decreasing precipitation. The annual maximum streamflow series of the Huangzhuang station have significantly changed in 1968, because the Danjiangkou Reservoir has been put into operation. Moreover, the annual maximum streamflow series of the Huangzhuang station have moderately changed in 1992 because of the decreasing precipitation.
2. The determination of time-varying characteristics concentrates on the location and scale parameters, whereas the shape parameter is assumed to be constant with time in all distributions. The GAMLSS method can adequately model the time-varying characteristics in the location and scale parameters of all marginal and joint distributions of two annual maximum streamflow series at Ankang and Huangzhuang stations, while the Copula function can effectively accomplish the point estimations of non-stationary design floods. From the first time segment (1954–1967) to the next two time segments (1968–1991 and 1992–2018), the design flood values of the Ankang and Huangzhuang stations dramatically decrease under the non-stationary condition as compared with the stationary condition.
3. The BUP post-processing technique not only can effectively quantify the uncertainty of probabilistic interval estimation of non-stationary design floods, but also can reduce the probabilistic distribution to a small range. The results of this study highlight the benefits of non-stationary hydrological frequency analysis to planning and design of hydraulic engineering projects meanwhile water resources management.

We believe that improving the reliability and generalizability of flood frequency analysis and decreasing the uncertainty of probabilistic interval estimation of design floods would enhance the trust in non-stationary frequency analysis approaches and promote more practices in hydrological sciences.

ACKNOWLEDGEMENTS

This work was supported by the National Natural Science Foundation of China (No. U20A20317 and 51538173), the Research Council of Norway (FRINATEK Project 274310) and the China Three Gorges Corporation (No. 0799254). The authors thank the Editors and anonymous Reviewers for their constructive comments that greatly contributed to enrich the manuscript.

DATA AVAILABILITY STATEMENT

Data cannot be made publicly available; readers should contact the corresponding author for details.

REFERENCES

Agilan, V. & Umamahesh, N. V. 2017 What are the best covariates for developing non-stationary rainfall intensity-duration-frequency relationship? *Advances in Water Resources* **101**, 11–22.

Akaike, H. 1974 A new look at the statistical model identification. *IEEE Transactions on Automatic Control* **19** (6), 716–723.

Bayazit, M. 2015 Nonstationarity of hydrological records and recent trends in trend analysis: a state-of-the-art review. *Environmental Processes* **2** (3), 527–542.

Bian, G., Du, J., Song, M., Zhang, X., Zhang, X., Li, R., Wu, S., Duan, Z. & Xu, C. Y. 2020 Detection and attribution of flood responses to precipitation change and urbanization: a case study in Qinhuai River Basin, Southeast China. *Hydrology Research* **51** (2), 351–365.

Bracken, C., Holman, K. D., Rajagopalan, B. & Moradkhani, H. 2018 A Bayesian hierarchical approach to multivariate nonstationary hydrologic frequency analysis. *Water Resources Research* **54** (1), 243–255.

Chebana, F., Ouarda, T. B. M. J. & Duong, T. C. 2013 Testing for multivariate trends in hydrologic frequency analysis. *Journal of Hydrology* **486** (486), 519–530.

Chen, L. & Guo, S. 2018 *Copulas and its Application in Hydrology and Water Resources*. Springer Verlag, Singapore.

Cheng, L. & AghaKouchak, A. 2015 Nonstationary precipitation intensity-duration-frequency curves for infrastructure design in a changing climate. *Scientific Reports* **4** (1), 7093–7093.

Fan, Y. R., Huang, W. W., Huang, G. H., Li, Y. P., Huang, K. & Li, Z. 2016 Hydrologic risk analysis in the Yangtze river basin through coupling Gaussian mixtures into copulas. *Advances in Water Resources* **88**, 170–185.

Gilroy, K. L. & McCuen, R. H. 2012 A nonstationary flood frequency analysis method to adjust for future climate change and urbanization. *Journal of Hydrology* **414**, 40–48.

Hui, R., Herman, J., Lund, J. & Madani, K. 2018 Adaptive water infrastructure planning for nonstationary hydrology. *Advances in Water Resources* **118**, 83–94.

James, N. A. & Matteson, D. S. 2015 Ecp: an R package for nonparametric multiple change point analysis of multivariate data. *Journal of Statistical Software* **62** (1), 1–25.

Jiang, C., Xiong, L., Xu, C.-Y. & Guo, S. 2015 Bivariate frequency analysis of nonstationary low-flow series based on the time-varying copula. *Hydrological Processes* **29** (6), 1521–1534.

Joe, H. 1997 *Multivariate Models and Multivariate Dependence Concepts*. Chapman and Hall, London.

Krzysztofowicz, R. 1999 Bayesian theory of probabilistic forecasting via deterministic hydrologic model. *Water Resources Research* **35** (9), 2739–2750.

Kwon, H.-H. & Lall, U. 2016 A copula-based nonstationary frequency analysis for the 2012–2015 drought in California. *Water Resources Research* **52** (7), 5662–5675.

Li, J., Liu, X. & Chen, F. 2015 Evaluation of nonstationarity in annual maximum flood series and the associations with large-scale climate patterns and human activities. *Water Resources Management* **29** (5), 1653–1668.

Li, J., Lei, Y., Tan, S., Bell, C. D., Engel, B. A. & Wang, Y. 2018 Nonstationary flood frequency analysis for annual flood peak and volume series in both univariate and bivariate domain. *Water Resources Management* **32** (13), 4239–4252.

Li, Z. & Krafty, R. T. 2019 Adaptive Bayesian time–frequency analysis of multivariate time series. *Journal of the American Statistical Association* **114** (525), 453–465.

Li, H., Wang, D., Singh, V. P., Wang, Y., Wu, J., Wu, J., Liu, J., Zou, Y. & He, R. 2019 Non-stationary frequency analysis of annual extreme rainfall volume and intensity using Archimedean copulas: a case study in eastern China. *Journal of Hydrology* **571**, 114–131.

Li, S., Qin, Y., Liu, Y., Song, X., Liu, Q. & Li, Z. 2020 Estimating the design flood under the influence of check dams by removing nonstationarity from the flood peak discharge series. *Hydrology Research* **51** (6), 1261–1273.

Liang, Z., Yang, J., Hu, Y., Wang, J., Li, B. & Zhao, J. 2018 A sample reconstruction method based on a modified reservoir index for flood frequency analysis of non-stationary hydrological series. *Stochastic Environmental Research and Risk Assessment* **32** (6), 1561–1571.

Liu, D., Wang, D., Singh, V. P., Wang, Y., Wu, J., Wang, L., Zou, X., Chen, Y. & Chen, X. 2017 Optimal moment determination in POME-copula based hydrometeorological dependence modelling. *Advances in Water Resources* **105**, 39–50.

Liu, S., Huang, S., Xie, Y., Wang, H., Leng, G., Huang, Q., Wei, X. & Wang, L. 2019 Identification of the non-stationarity of floods: changing patterns, causes, and implications. *Water Resources Management* **33** (3), 939–953.

López, J. & Francés, F. 2013 Non-stationary flood frequency analysis in continental Spanish rivers, using climate and reservoir indices as external covariates. *Hydrology and Earth System Sciences* **17** (8), 3189–3203.

Ngongondo, C., Zhou, Y. & Xu, C.-Y. 2020 Multivariate framework for the assessment of key forcing to Lake Malawi level variations in non-stationary frequency analysis. *Environmental Monitoring and Assessment* **192** (9), 1–23.

Parent, E., Favre, A.-C., Bernier, J. & Perreault, L. 2014 Copula models for frequency analysis what can be learned from a Bayesian perspective. *Advances in Water Resources* **63**, 91–103.

Pettitt, A. N. 1979 A non-parametric approach to the change-point problem. *Journal of The Royal Statistical Society Series C (Applied Statistics)* **28** (2), 126–135.

Ragno, E., AghaKouchak, A., Cheng, L. & Sadegh, M. 2019 A generalized framework for process-informed nonstationary extreme value analysis. *Advances in Water Resources* **130**, 270–282.

Read, L. K. & Vogel, R. M. 2015 Reliability, return periods, and risk under nonstationarity. *Water Resources Research* **51** (8), 6381–6398.

Reeves, J., Chen, J., Wang, X. L., Lund, R. & Lu, Q. Q. 2007 A review and comparison of changepoint detection techniques for climate data. *Journal of Applied Meteorology and Climatology* **46** (6), 900–915.

Rougé, C., Ge, Y. & Cai, X. 2013 Detecting gradual and abrupt changes in hydrological records. *Advances in Water Resources* **53**, 33–44.

Sarhadi, A., Burn, D. H., Ausín, M. C. & Wiper, M. P. 2016 Time-varying nonstationary multivariate risk analysis using a dynamic Bayesian copula. *Water Resources Research* **52** (3), 2327–2349.

Şen, Z. 2012 Innovative trend analysis methodology. *Journal of Hydrologic Engineering* **17** (9), 1042–1046.

Serago, J. M. & Vogel, R. M. 2018 Parsimonious nonstationary flood frequency analysis. *Advances in Water Resources* **112**, 1–16.

Song, M. M., Zhang, J. Y., Bian, G. D., Wang, J. & Wang, G. Q. 2020 Quantifying effects of urban land-use patterns on flood regimes for a typical urbanized basin in eastern China. *Hydrology Research* **51** (6), 1521–1536.

Stasinopoulos, D. M. & Rigby, R. A. 2007 Generalized additive models for location scale and shape (GAMLSS) in R. *Journal of Statistical Software* **23** (1), 1–46.

Stasinopoulos, D. M., Rigby, R. A. & Akantziliotou, C. 2008 *Instructions on How to Use the GAMLSS Package in R Second Edition*. London. Available from: http://www.gamlss.org.

Su, C. & Chen, X. 2019 Assessing the effects of reservoirs on extreme flows using nonstationary flood frequency models with the modified reservoir index as a covariate. *Advances in Water Resources* **124**, 29–40.

Sun, X., Li, Z. & Tian, Q. 2020 Assessment of hydrological drought based on nonstationary runoff data. *Hydrology Research* **51** (5), 894–910.

Tan, Q. F., Mao, Y. Z., Wen, X., Jin, T., Ding, Z. Y. & Wang, Z. N. 2021 Copula-based modeling of hydraulic structures using a nonlinear reservoir model. *Hydrology Research*. https://doi.org/10.2166/nh.2021.178.

Thorarinsdottir, T. L., Hellton, K. H., Steinbakk, G. H., Schlichting, L. & Engeland, K. 2018 Bayesian regional flood frequency analysis for large catchments. *Water Resources Research* **54** (9), 6929–6947.

Uranchimeg, S., Kwon, H.-H., Kim, B. & Kim, T.-W. 2020 Changes in extreme rainfall and its implications for design rainfall using a Bayesian quantile regression approach. *Hydrology Research* **51** (4), 699–719.

Villarini, G. & Smith, J. A. 2010 Flood peak distributions for the eastern United States. *Water Resources Research* **46** (6), 1–17.

Villarini, G. & Strong, A. 2014 Roles of climate and agricultural practices in discharge changes in an agricultural watershed in Iowa. *Agriculture, Ecosystems & Environment* **188**, 204–211.

Villarini, G., Serinaldi, F., Smith, J. A. & Krajewski, W. F. 2009 On the stationarity of annual flood peaks in the continental United States during the 20th century. *Water Resources Research* **45** (8), 1–17.

Villarini, G., Smith, J. A., Serinaldi, F., Ntelekos, A. A. & Schwarz, U. 2012 Analyses of extreme flooding in Austria over the period 1951–2006. *International Journal of Climatology* **32** (8), 1178–1192.

Vinnarasi, R. & Dhanya, C. T. 2019 Bringing realism into a dynamic copula-based non-stationary intensity-duration model. *Advances in Water Resources* **130**, 325–338.

Wei, T. & Song, S. 2018 Copula-based composite likelihood approach for frequency analysis of short annual precipitation records. *Hydrology Research* **49** (5), 1498–1512.

Wu, X., Wang, Z., Zhou, X., Zeng, Z., Lai, C. & Chen, X. 2017 Variability of annual peak flows in the Beijiang River Basin, South China, and possible underlying causes. *Hydrology Research* **48** (2), 442–454.

Xiong, L., Yu, K. & Gottschalk, L. 2014 Estimation of the distribution of annual runoff from climatic variables using copulas. *Water Resources Research* **50** (9), 7134–7152.

Xiong, L., Jiang, C., Xu, C., Yu, K. & Guo, S. 2015a A framework of change-point detection for multivariate hydrological series. *Water Resources Research* **51** (10), 8198–8217.

Xiong, L., Du, T., Xu, C.-Y., Guo, S., Jiang, C. & Gippel, C. J. 2015b Non-stationary annual maximum flood frequency analysis using the norming constants method to consider non-stationarity in the annual daily flow series. *Water Resources Management* **29** (10), 3615–3633.

Xiong, B., Xiong, L., Chen, J., Xu, C.-Y. & Li, L. 2017 Multiple causes of nonstationarity in the Weihe annual low-flow series. *Hydrology and Earth System Sciences* **22** (2), 1525–1542.

Xu, W., Jiang, C., Yan, L., Li, L. & Liu, S. 2018 An adaptive metropolis-hastings optimization algorithm of Bayesian estimation in non-stationary flood frequency analysis. *Water Resources Management* **32** (4), 1343–1366.

Yan, L., Xiong, L., Guo, S., Xu, C.-Y., Xia, J. & Du, T. 2017a Comparison of four nonstationary hydrologic design methods for changing environment. *Journal of Hydrology* **551**, 132–150.

Yan, L., Xiong, L., Liu, D., Hu, T. & Xu, C. 2017b Frequency analysis of nonstationary annual maximum flood series using the time-varying two-component mixture distributions. *Hydrological Processes* **31** (1), 69–89.

Yan, L., Xiong, L., Ruan, G., Xu, C.-Y., Yan, P. & Liu, P. 2019 Reducing uncertainty of design floods of two-component mixture distributions by utilizing flood timescale to classify flood types in seasonally snow covered region. *Journal of Hydrology* **574**, 588–608.

Yan, L., Xiong, L., Ruan, G., Xu, C.-Y. & Zhang, M. J. 2021 Design flood estimation with varying record lengths in Norway under stationarity and nonstationarity scenarios. *Hydrology Research*. https://doi.org/10.2166/nh.2021.026.

Yu, K., Xiong, L. & Gottschalk, L. 2014 Derivation of low flow distribution functions using copulas. *Journal of Hydrology* **508**, 273–288.

Yu, K., Gottschalk, L., Zhang, X., Li, P., Li, Z., Xiong, L. & Sun, Q. 2018 Analysis of nonstationarity in low flow in the Loess Plateau of China. *Hydrological Processes* **32** (12), 1844–1857.

Yue, S., Pilon, P., Phinney, B. & Cavadias, G. 2002 The influence of autocorrelation on the ability to detect trend in hydrological series. *Hydrological Processes* **16** (9), 1807–1829.

Zhang, D., Yan, D., Wang, Y.-C., Lu, F. & Liu, S. 2015 GAMLSS-based nonstationary modeling of extreme precipitation in Beijing–Tianjin–Hebei region of China. *Natural Hazards* **77** (2), 1037–1053.

Zhang, L. & Singh, V. P. 2019 *Copulas and Their Applications in Water Resources Engineering*. Cambridge University Press, Cambridge.

Zhou, Y., Guo, S., Hong, X. & Chang, F.-J. 2017 Systematic impact assessment on inter-basin water transfer projects of the Hanjiang River Basin in China. *Journal of Hydrology* **553**, 584–595.

Zhou, Y. 2020 Exploring multidecadal changes in climate and reservoir storage for assessing nonstationarity in flood peaks and risks worldwide by an integrated frequency analysis approach. *Water Research* **185**, 116265.

First received 11 January 2021; accepted in revised form 18 November 2021. Available online 8 December 2021

doi: 10.2166/nh.2021.006

Propagation dynamics and causes of hydrological drought in response to meteorological drought at seasonal timescales

Lan Ma[a], Qiang Huang[a], Shengzhi Huang[a,*], Dengfeng Liu[a], Guoyong Leng[b], Lu Wang[a] and Pei Li[a]

[a] State Key Laboratory of Eco-hydraulics in Northwest Arid Region of China, Xi'an University of Technology, Xi'an 710048, China
[b] Key Laboratory of Water Cycle and Related Land Surface Processes, Institute of Geographic Sciences and Natural Resources Research, Chinese Academy of Sciences, Beijing 100101, China
*Corresponding author. E-mail: huangshengzhi7788@126.com

ABSTRACT

According to the widely accepted definition of drought, meteorological and hydrological droughts originally develop from rainfall and runoff deficits, respectively. Runoff deficit is mainly derived from rainfall deficit, and the propagation from meteorological drought to hydrological drought is critical for agricultural water management. Nevertheless, the characteristics and dynamics of drought propagation in the spatio-temporal scale remain unresolved. To this end, the characteristics and dynamics of drought propagation in different seasons and their linkages with key forcing factors are evaluated. In this study, meteorological and hydrological droughts are characterized by the Standardized Precipitation Index (SPI) and the Standardized Runoff Index (SRI), respectively. Propagation time is identified by the corresponding timescale of the maximum correlation coefficient between the SPI and the SRI. Then, a 20-year sliding window is adopted to explore the propagation dynamic in various seasons. Furthermore, the multiple linear regression model is established to quantitatively explore the influence of meteorological factors, underlying surface features and teleconnection factors on the propagation time variations. The Wei River Basin, a typical Loess Plateau watershed in China, is selected as a case study. Results indicate the following: (1) the propagation time from meteorological to hydrological drought is shorter in summer (2 months) and autumn (3 months), whereas it is longer in spring (8 months) and winter (13 months). Moreover, the propagation rates exhibit a decreasing trend in warm seasons, which, however, show an increasing trend in cold seasons; (2) a significant slowing propagation in autumn is mainly caused by the decreasing soil moisture and precipitation, whereas the non-significant tendency in summer is generally induced by the offset between insignificant increasing precipitation and significant decreasing soil moisture; (3) the replenishment from streamflow to groundwater in advance prompts the faster propagation from meteorological to hydrological drought in spring and winter and (4) teleconnection factors have strong influences on the propagation in autumn, in which Arctic Oscillation, El Niño-Southern Oscillation and Pacific Decadal Oscillation mainly affect participation, arid index and soil moisture, thereby impacting drought propagation.

Key words: drought propagation dynamics, SPI, SRI, teleconnection factors

HIGHLIGHTS

- The propagation dynamics at a seasonal timescale were explored.
- The impacts of diverse factors on the propagation dynamics were investigated. Autumn exhibits a significant increasing propagation time mainly induced by decreasing soil moisture and precipitation.
- Winter shows a significant decreasing propagation time mainly caused by earlier groundwater supply.
- Teleconnection factors exert strong influences on the propagation process in autumn.

GRAPHICAL ABSTRACT

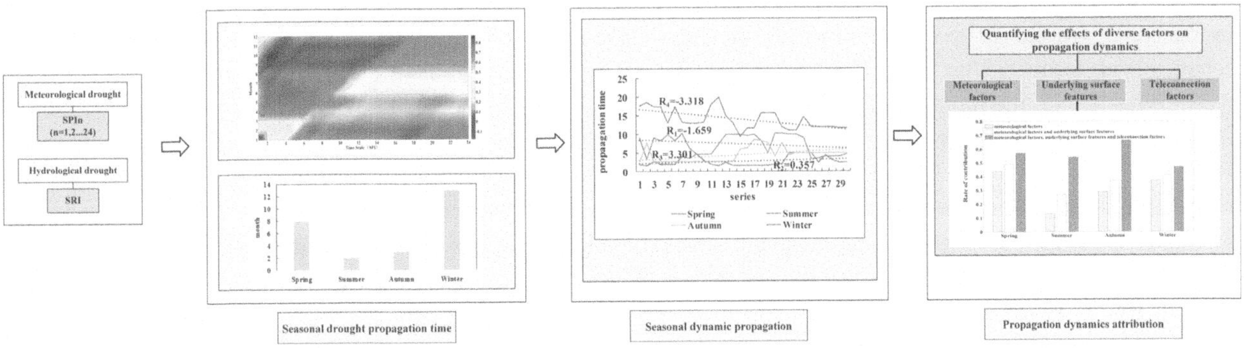

1. INTRODUCTION

With global climate change and intensified water circulation, extreme weather and climate events have become more frequent in recent years. History has repeatedly proved that serious drought events in a given area will not only cause great mortality and economic losses, but also arouse the widespread concern of relevant scholars as well as the introspection of local government. Numerous drought and flood events have occurred in recent decades. A string of drought events occurred in recent decades, thereinto, the sandstorm drought in America of the 1930s (Brekke 1947; Wilhite & Glant 1985; Cook *et al.* 1999; Fye *et al.* 2010), which covered across two-thirds of the United States and wreaking havoc for a decade, caused 250 million people to leave. China has been subjected to drought damage repeatedly in the past years. The year of 2010 in the southwestern of China affected >67,000 km^2 of arable land and left >20 million people without drinking water. The middle and lower reaches of the Yangtze River have suffered the worst drought for a 70-year return period (Yu *et al.* 2014), which impacted 17 million people and the direct economic losses reached nearly 10 billion.

Drought as a natural hazard may happen in all the places (including dry and wet regions) at multiple time scales (Fang *et al.* 2019a, 2019b). It is deemed as a complex, multifaceted phenomenon diffusely (Van Loon 2015), which has a major impact on every part of the hydrological cycle. Moreover, drought can influence wide areas and last for a long time (WMO 1997). Drought has attracted growing attention in many different fields, such as ecology, environment, hydrology and agriculture (Tallaksen & Van Lanen 2004; Mishra & Singh 2010; Zhang *et al.* 2012; Mohammad *et al.* 2014; Chen & Sun 2017). Drought is largely driven by climate changes, such as the decreases of precipitation, which develops slowly and lasts months to years (Tallaksen & Van Lanen 2004; Tallaksen *et al.* 2009). Depending on the deficit of different hydrologic features, drought is commonly divided into four categories, that is, meteorological drought, hydrological drought, agricultural drought and socioeconomic drought (Mishra & Singh 2010). There are some differences and connections between them. Specifically, when the precipitation is insufficient, meteorological drought occurs; the lack of runoff then causes hydrological drought; agricultural drought is triggered by the inadequacy of soil moisture; whereas the socioeconomic drought directly results from the water supply not satisfying the human water demand, which is also the embodiment of the impacts of the above three types of droughts on human society (Dracup *et al.* 1980; Guo *et al.* 2019). In general, meteorological drought begins the earliest, then agricultural and hydrological droughts happen. If the socioeconomic drought occurs finally, this water deficit event is more serious. Many researches have focused on the establishment of drought index, to identify drought events and evaluate spatial and temporal distribution characteristics of drought (Feng *et al.* 2014; Huang *et al.* 2015; Leng *et al.* 2015; Xu *et al.* 2015; Ge *et al.* 2016; Huang *et al.* 2019).

As mentioned above, there are close links between different types of drought. The transformation from one type of drought to another is called drought propagation (Huang *et al.* 2017). In recent years, drought propagation has become a hot topic in the field of hydrology and water resources. The analysis methods of drought propagation are mainly divided into model simulations and statistical methods. Different hydrological models reveal the propagation time between different types of droughts to a certain extent. Wang *et al.* (2011) and Van Loon & Van Lanen (2012) adopted a hydro-agricultural coupling model and an integrated large-scale hydrological model to study the propagation of drought. Lorenzo-Lacruz *et al.* (2013) used the Pearson correlation coefficient to determine the time scale of the Standardized Precipitation Index

(SPI) with the best correlation with the Standardized Runoff Index (SRI) as the response time of hydrological drought to cumulative precipitation deficit. Besides, some researches discuss the forcing factors on drought propagation, for example, the process of propagation from meteorological to hydrological drought is influenced by climate characteristics (Van Loon & Van Lanen 2012). Furthermore, underlying surface features and teleconnection factors also have a great influence on the propagation process (Huang *et al.* 2015, 2017). For the law of drought propagation, Li *et al.* (2018) established the relationship between hydrological drought and driving factors, which can explain this pattern to some extent.

Although multiple researches have inquired into the relationships and influences between different types of drought, the dynamic change of drought propagation and the quantitative estimation of various influencing factors on its dynamics at seasonal scale remain unsolved. Improving the understanding of the mechanisms of drought propagation can help water resource managers to design optimal water management strategies, which may be used to mitigate the impacts of droughts. However, propagation in diverse types of drought needs further and systematic research. The numerous mechanisms behind drought propagation require to be explored in depth. To fill this research gap, the study attempts to examine the propagation time from meteorological to hydrological drought in different seasons, and analyzes their dynamic changes, for evaluating the drought propagation dynamics. With the modification of precipitation deficits in drought propagation, diverse climate, different basins as well as human activities make an important contribution, leading to the diversity of drought propagation in different catchments (Van Loon 2015). Thereupon, examining the different types of drought and how drought propagates is the top priority. It can boost the acquaintance of drought, create the foundation to support the development of drought and predict the variation tendency of drought.

This study has the following three main objectives: (1) to calculate the propagation time from meteorological drought to hydrological drought at seasonal scale; (2) to examine the seasonal dynamic propagation characteristics under the backdrop of a changing environment and (3) to quantitatively explore the impacts of meteorology, underlying surface features, teleconnection factors on seasonal dynamic propagation.

2. STUDY AREA AND DATASET

2.1. Study area

The Wei River Basin (WRB) is the largest tributary of the Yellow River basin, which covers an area of 134,600 km^2. The distribution map of the WRB is shown in Figure 1. This basin is located in the continental monsoon climate zone with abundant precipitation and high temperature in summer, little precipitation and low temperature in winter. Generally speaking, precipitation has a significant seasonal variation. During monsoon (June–September), precipitation accounts for 60% of whole year's precipitation, which always appears in the form of rainstorm or continuous rain. Owing to the uneven annual distribution and large inter-annual variation of precipitation in this area, which are mainly caused by the instable intensity of the North Pacific subtropics. The annual maximum precipitation can be 2.5 times larger than the annual minimum precipitation. Since ancient times, natural disaster events have frequently occurred in this basin (Du 2008), especially for drought events, which is the most frequent and harmful disaster. Drought has become the main factor restricting the regional development.

The Wei River is the main water source of the Guanzhong Plain, which is an area where agriculture plays a crucial role in local economic development (Huang *et al.* 2017). With the rapid development, this zone has also built many industrial parks, whose functioning requires water (Huang *et al.* 2017). The runoff in the WRB has obviously reduced in recent decades (Huang *et al.* 2014a, 2014b), and it was often subjected to the effect of droughts (Huang *et al.* 2014a, 2014b). Hence, the WRB is a region with extreme shortage of water. A comprehensive drought investigation integrated with the drought propagation and its influence is of great important in mitigating the drought effects.

2.2. Data

In this paper, daily precipitation data were collected from 1960 to 2010 from 21 weather stations in the WRB, which derived from the National Climate Center of the China Meteorological Administration. The runoff data were obtained from the Yellow River conservancy commission, including daily runoff sequences from 1962 to 2010 from several hydrological stations in Linjiacun, Zhangjiashan, Huaxian and Zhuangtou and others. The mean values of these stations are used as the calculated data of the WRB in this study. The El Niño-Southern Oscillation (ENSO) events are represented by the monthly Nino 3.4

Index and the monthly Arctic Oscillation (AO) series containing 1962–2010 events obtained from the National Oceanic and Atmospheric Administration (NOAA). Daily sunspot data from 1962 to 2010 supplied by the International Council for Science (ICSU) World Data System (WDS) and the long-term observational series were used in this study (http://sidc.oma.be/silso/dayssnplot). Additionally, potential evapotranspiration series of the meteorological stations were calculated through the Penman–Monteith equation (Allen *et al.* 1989). The soil moisture and base flow data were both simulated by the Variable Infiltration Capacity (VIC) model.

3. METHODOLOGIES

3.1. Standardized Precipitation Index

It is hard to compare changes in diverse spatial–temporal scale utilizing the precipitation data straightly due to the high fluctuation of precipitation in different times and districts. Moreover, the distribution of precipitation is a skewed distribution. On that basis, McKee *et al.* (1993) first brought up the conception of SPI to address a variety of time scales, which can monitor long-term precipitation conditions. The SPI can transform the skewness distribution to normal distribution, which is not only convenient for different data comparisons, but also can eliminate spatial and temporal distribution differences. At present, many studies have applied the SPI to characterize the meteorological arid conditions in the WRB and all have good results (Huang *et al.* 2014a, 2014b; Cai *et al.* 2018).

SPI is derived from a long-term record of precipitation, which is further transformed into standard normal distribution. In the first place, the original precipitation observations are aggregated to generate a precipitation series at a timescale of interest (typically 1, 3, 6 or 12 months). Then, the gamma probability distribution function is used to fit the average precipitation sequence. At last, the cumulative probability of the gamma distribution is converted to the standard normal distribution to calculate the SPI. The detail calculation process of SPI is introduced by Fang *et al.* (2019b).

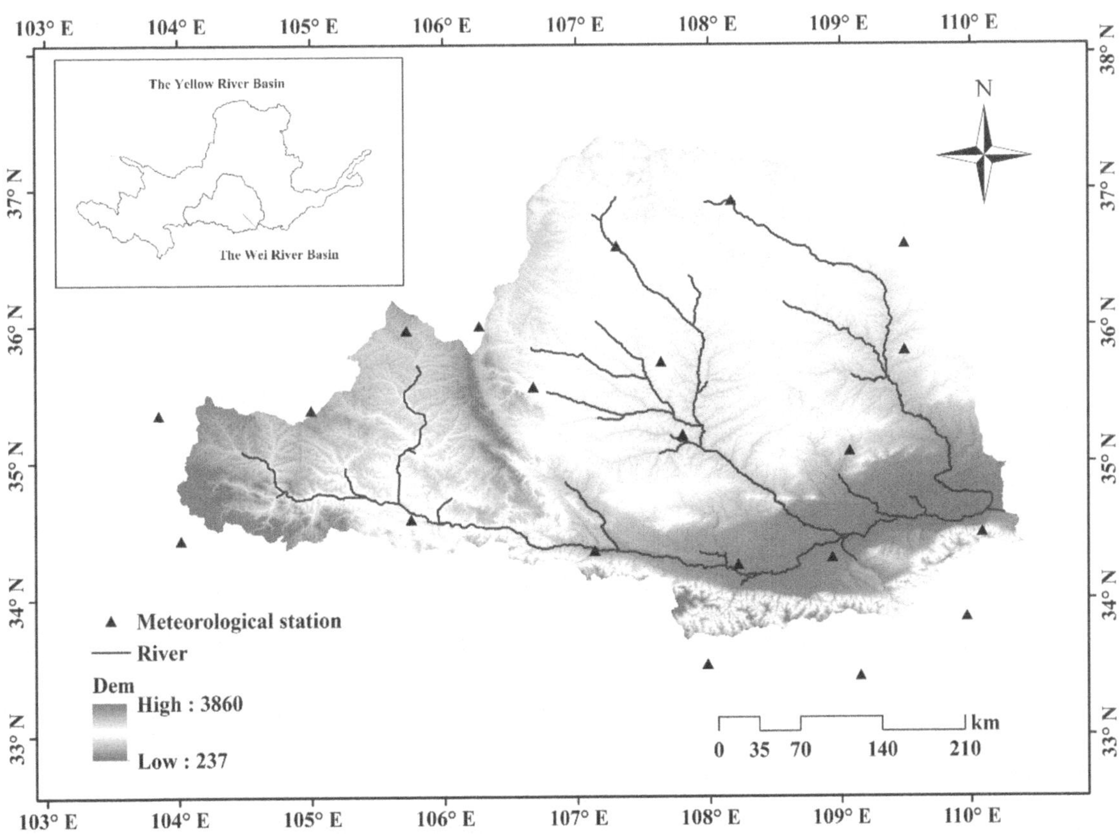

Figure 1 | Distribution map of the WRB belonging to the Yellow River Basin, China.

3.2. Standardized Runoff Index

The SRI was put forward on the strength of SPI theory used by McKee *et al.* (1993). Also, it was first proposed by Shukla & Wood (2008). Compared to the SPI, they take the same principles to build only using the different preliminary. That is to say, the calculation of SRI substitutes the runoff series for precipitation series of SPI. The calculation of SRI is to transform the distribution of accumulated runoff on a given time scale into a standard normal distribution through equal probability transformation.

3.3. Calculation of propagation time from meteorological to hydrological drought

In fact, the propagation from meteorological drought to hydrological drought is the conversion process from precipitation to runoff. Based on the multi-scale property of SPI, the scale range of SPI is changed, the scale range of SRI is fixed and then the correlation coefficient between the SPI in a variable scale and the SRI can be calculated. The SPI is chosen at the scale of 1–24 months and the SRI at 1 month in this study. Then, the correlations of SPI sequences at different time scales and SRI sequences are all calculated, respectively. Finally, the SPI scale corresponding to the maximum correlation coefficient is regarded as the propagation time from meteorological to hydrological drought (Yu *et al.* 2018; Wang *et al.* 2019).

The greatest advantage of the SPI is that it can characterize drought conditions at different scales, which would be perfectly applied on not only short-term, but also long-term, water resource monitoring. To investigate the propagation time from meteorological to hydrological drought, the relationship between the SRI and the SPI at different time scales in the WRB is explored, and related results are displayed in Figure 2. In the figure, the *x* coordinate signifies various time scales of SPI (1, 2, 3, …, 24), while the *y* coordinate displays months. Different colors imply the correlation coefficients of SPI and SRI, where red represents the high value region and blue represents the low value region of the correlation coefficient.

3.4. The cross-wavelet

The cross-wavelet is capable of presenting the correlation between two time series in both time and frequency domains, combined with the wavelet transformation and cross-spectrum analysis (Hudgins & Huang 1996). In the figure, the red represents the peak value of energy density, and the blue represents the valley value of energy density, respectively, showing the locality and dynamic characteristics of time-sharing transformation of the dominant wave group. The dark or light color represents the relative change of energy density. The fine solid black line represents the boundary of the cone line affected by the wavelet, and its interior is an effective spectral value region. The thick black line in the region represents a 95% confidence interval with a passing significance level of $A = 0.05$. The direction of the arrows in the figure reflects the phase relationship between

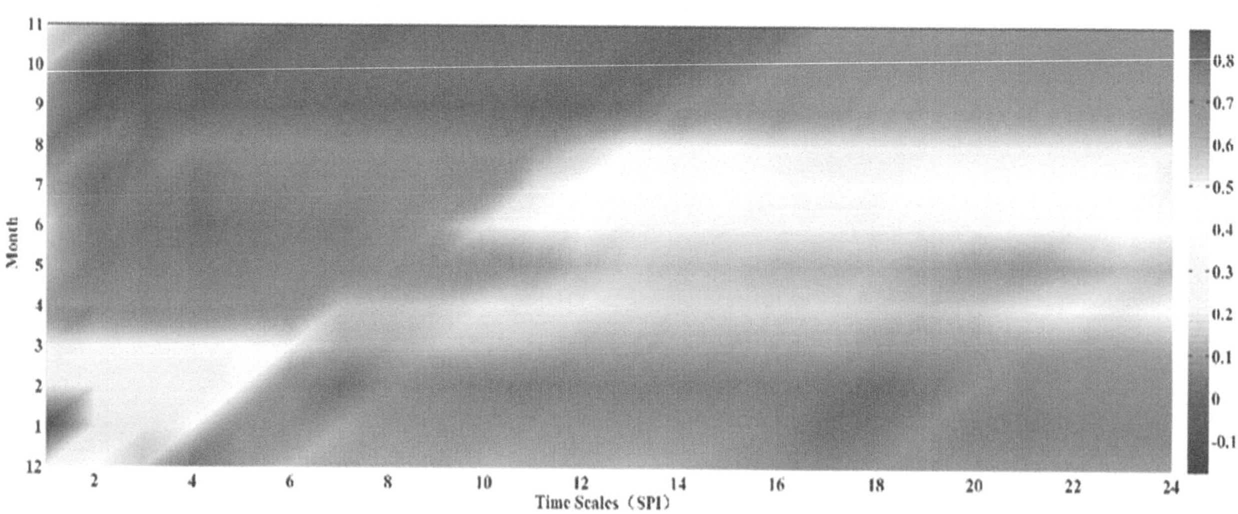

Figure 2 | Relationship between the 1-month SRI and the SPI at different time scales in the WRB. Please refer to the online version of this paper to see this figure in colour: http://dx.doi.10.2166/nh.2021.006.

the influencing factors and rainfall erosivity. The (→) from left to right indicates that the changes in the two factors are in the same phase, showing a positive correlation. The (←) from the right finger to the left represents the inverse phase, showing a negative correlation (Grinsted *et al.* 2004).

3.5. Multiple linear regression model

Multiple linear regression (MLR) is a statistical technique that attempts to model the relationship between two or more explanatory variables and a response variable by fitting a linear equation to observed data. Every value of the independent variable x is associated with a value of the dependent variable y. Formally, the model for MLR, given n observations, is

$$yi = \beta_0 + \beta_1 xi1 + \beta_2 xi2 + \cdots + \beta_p xip + c \ (i = 1, 2, \ldots, n) \quad (1)$$

MLR is used to determine a mathematical relationship among a number of random variables. In other terms, MLR examines how multiple independent variables are related to one dependent variable. Once each of the independent factors have been determined to predict the dependent variable, the information on the multiple variables can be used to create an accurate prediction on the level of effect they have on the outcome variable. The model creates a relationship in the form of a straight line (linear) that best approximates all the individual data points.

4. RESULTS AND DISCUSSION

4.1. Propagation time from meteorological to hydrological drought in different seasons

The relationship of the SPI and the SRI between 1962 and 2010 in the WRB was explored. The outcome reflects the response time of hydrological drought to meteorological drought is roughly 3 months corresponding to the coefficient of 0.870. The propagation time is short in summer and autumn, but relatively long in spring and winter. In terms of the correlation coefficient of all months, the maximum appears in October, and the autumn has the strongest correlation.

Generally, the WRB is strongly affected by monsoon, which has abundant precipitation and high temperature in summer, dry and cold climate occurs in winter. The annual average temperature is about 6–14 °C, and the annual average precipitation is 450–700 mm (Huang *et al.* 2015).

Summer (June–August) and autumn (September–November) are wet seasons. Temperature is relatively high in these two seasons, especially in summer. Its average temperature reaches 24 °C. Precipitation is always abundant (average 161.6 mm in autumn and 286.4 mm in summer in 1962–2010) with high-intensity rainstorms (Table 1). Once precipitation occurs, it is easy to generate runoff since soil moisture is high. After precipitation replenishes the interception, infiltration and sink filling, the loss is no longer significantly increased with the continuation of precipitation, and the soil is basically saturated, resulting in the generation of runoff (Apurv *et al.* 2017; Joo *et al.* 2020). The amount and intensity of the precipitation in summer and autumn are large. Therefore, the time of drought propagation is always short in the two seasons. Table 1 shows that the amount of the precipitation in summer is more than autumn, and land surface converge, therefore, tends to be faster. As a result, the propagation time in summer is shorter than that in autumn. It is indicated in Table 1 that spring (March–May) and winter (December–February) are dry seasons with little precipitation, especially in winter (120.6 mm in spring and 18.0 mm in winter in 1962–2010), and temperature is relatively low in these two seasons, especially in winter, and its average temperature is −1 to −3 °C (Jiang *et al.* 2012). Almost all of the runoff is supplied by groundwater discharge in the form of base flow in dry seasons, which is generally derived from long-term precipitation accumulation of early period (Apurv *et al.* 2017). It is found that the runoff in this region mainly depends on groundwater formed by the accumulation of precipitation in the previous 8 months in spring and 13 months in winter. The slow development of water cycle results in few runoff generation. In consequence, the process of drought propagation tends to

Table 1 | Propagation time from meteorological to hydrological drought and related meteorology information

Statistic	Spring	Summer	Autumn	Winter
P (mm)	120.6	286.4	161.6	18.0
T (month)	8	2	3	13

be slow. Moreover, the propagation time in winter is longer than that in spring mainly caused by less precipitation and cold weather. The runoff in spring comes from both groundwater and direct surface runoff, thus leading to faster propagation than winter.

4.2. Seasonal dynamic propagation characteristics from meteorological to hydrological drought

Figure 3 shows the trends of propagation time in different seasons, which is derived from a sliding window with 20-year length between 1962 and 2010. Then, the Mann–Kendall test was used to verify the significance of these trends. Specifically, Table 2 indicates that the propagation time has a significant decreasing trend in winter and a significant increasing trend in autumn at the 99% confidence level. The propagation time has no obvious tendency in spring and summer. Results indicate that the hydrologic circle rate is intensifying in winter and reducing in autumn in the Loess Plateau watershed under the warming climate. The possible causes of the propagation dynamics are systematically investigated in the next section.

4.3. Effects of diverse influencing factors on seasonal dynamic propagation

Table 2 shows the correlation between influencing factors and propagation time in different seasons. The tendency of meteorological factors and underlying surface features is shown in Table 3. It can be found that the arid index also has an insignificant decreasing trend, while precipitation has an insignificant increasing trend in summer (Table 3). It indicates that the region in summer is wetter than before, which tends to make the propagation process faster. The significant decreasing trends of base flow and soil moisture require more water to replenish them first, thus making the propagation process slower. Besides, base flow and soil moisture all have better correlation with propagation time than precipitation and arid index. As a consequence, the interaction of insignificant increasing precipitation and significant decreasing soil moisture eventually results in the insignificant increasing propagation time in summer. It indicates that underlying surface features are major influencing factors.

Results in Table 3 also indicated that the precipitation has a significant downtrend at the 95% confidence level, and the arid index has a significant uptrend at the 95% confidence level in autumn. These two factors lead to the basin becoming drier, and the significant decreasing precipitation causes the decrease of runoff, thus making the propagation process slower. In

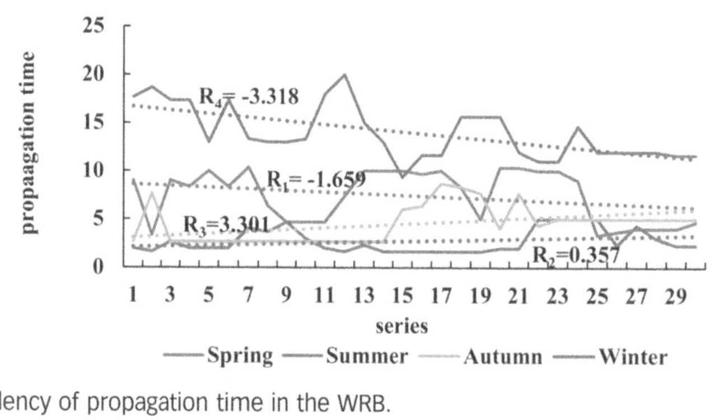

Figure 3 | Seasonal variation tendency of propagation time in the WRB.

Table 2 | Correlation between propagation time and meteorological factors, underlying surface features and teleconnection factors

	Meteorological factors			Underlying surface features		Teleconnection factors			
CC	P	ET_0	AI	SM	BF	SP	ENSO	PDO	AO
Spring	−0.598**	−0.093	0.561**	0.329*	0.261	0.474**	0.0182	−0.439**	−0.322*
Summer	−0.251	0.322*	0.296*	−0.318*	−0.315*	−0.596**	0.0177	0.238	0.227
Autumn	0.245	−0.467**	−0.368*	−0.176	−0.064	0.201	0.752**	0.560**	0.674**
Winter	0.604**	−0.567**	−0.601**	0.598**	0.617**	0.260	−0.556**	−0.515**	−0.200

CC, correlation coefficients; ET_0, potential evapotranspiration; P, precipitation; AI, aridity index; SP, sunspots; SM, soil moisture; BF, base flow.
*Significant at the 95% confidence level.
**Significant at the 99% confidence level.

Table 3 | Tendency of meteorological factors and underlying surface features

	Meteorological factors			Underlying surface features	
Influencing factors	**P**	**ET_0**	**AI**	**BF**	**SM**
Spring	−2.043*	1.715	1.871	−3.178**	−5.872**
Summer	0.422	−2.447*	−0.888	−2.328*	−3.440**
Autumn	−2.301*	0.089	2.060*	−1.635	−4.630**
Winter	1.543	−0.642	−1.612	−1.138	−4.473**

*Significant at the 95% confidence level.
**Significant at the 99% confidence level.

addition, base flow and soil moisture significantly decrease, especially for soil moisture. The replenishment of soil moisture needs more time, and the propagation time prolongs. As a result, both significant decreasing precipitation and soil moisture eventually lead to the significant increasing propagation time in autumn. This is the reason why the propagation time is more significantly increasing in autumn than that in summer.

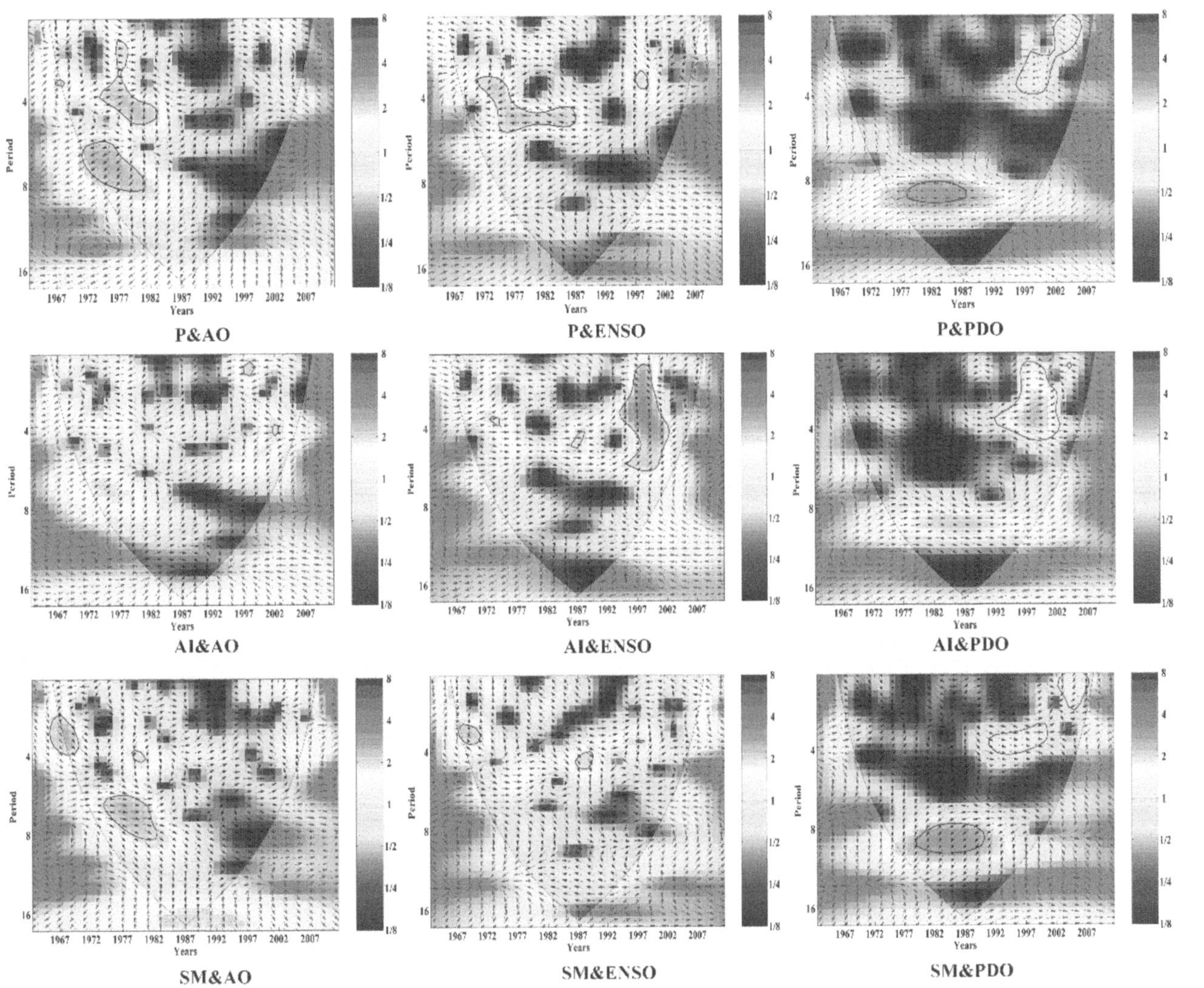

Figure 4 | Cross-wavelet of meteorological drought, underlying surface features and teleconnection factors exceeding the significance level in autumn.

Similarly, as noted in Table 3, a small amount of precipitation and high evapotranspiration cause scarce runoff generation, and the basin gets drier in winter. Streamflow in winter is reliant on groundwater supply, which is closely associated with the surface streamflow condition. Decreasing soil moisture and base flow tend to make surface streamflow exhaust earlier, thus leading to earlier replenishment of groundwater. Thus, the propagation time in winter has a significant decreasing trend. With regard to spring, its runoff is mainly impacted by precipitation and groundwater, where groundwater is the dominant factor due to the long propagation. Being similar with winter, decreasing soil moisture and base flow are expected to make surface streamflow exhaust earlier, thus resulting in earlier replenishment of groundwater. This leads to the faster propagation time. However, precipitation exhibits a significant downward trend at the 95% confidence level in spring, which causes a slower propagation time in the form of direct surface streamflow. Influenced by decreasing precipitation and earlier groundwater supply, the propagation time shows an insignificant decreasing trend.

The above results indicate that both meteorological factors and underlying surface features have a great impact of the dynamic changes of propagation time in autumn and winter. However, since the teleconnection factors can indirectly affect the propagation of drought, the study selects relatively significant influencing factors to explore the relationship between teleconnection factors and meteorological factors, as well as underlying surface features in these two seasons.

On the whole, the teleconnection factors have great influences on meteorological factors and underlying surface features. Specifically, the influence and correlation of different factors are different in autumn and winter.

The cross-wavelet analysis shown in Figure 4 indicated that the precipitation in autumn shows a significant positive correlation of about 6–8 years in 1972–1980 and a significant negative correlation of about 4–6 years in 1975–1984 with AO; the precipitation and ENSO have a significant negative correlation of about 4–6 years in 1970–1987; the precipitation and PDO have a significant negative correlation of about 9 years in 1979–1987, a significant negative correlation of about 2–4 years in 1997–2002 and a significant positive correlation about 2 years in 2002–2005. The arid index and AO have only a significant positive correlation of about 2 years in 1997–2000; the arid index and ENSO have a significant negative correlation of about 2–4 years in 1995–2000 and a significant positive correlation of about 4–6 years in 1995–2000; the arid index and PDO have a significant positive correlation of about 2–4 years in 1992–2002. The soil moisture and AO have a significant negative correlation of about 2–4 years in 1967–1970 and a significant positive correlation about 6–8 years in 1975–1982. The soil moisture and ENSO have a significant negative correlation of about 3 years in 1967–1972 and a significant negative correlation of about 4 years in 1987–1990. The soil moisture and PDO have a significant negative correlation of about 8 years in 1979–1990, a significant positive correlation of about 4 years in 1992–2002 and a significant negative

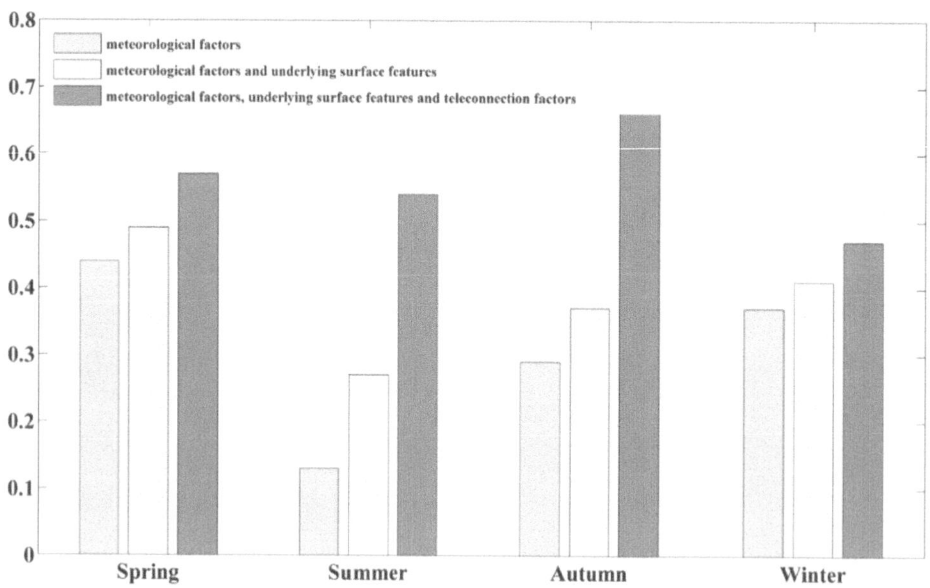

Figure 5 | Determination coefficient (R^2) and significance of the multiple linear regression model between propagation time and influencing factors. '*' represents significance at the 95% confidence level.

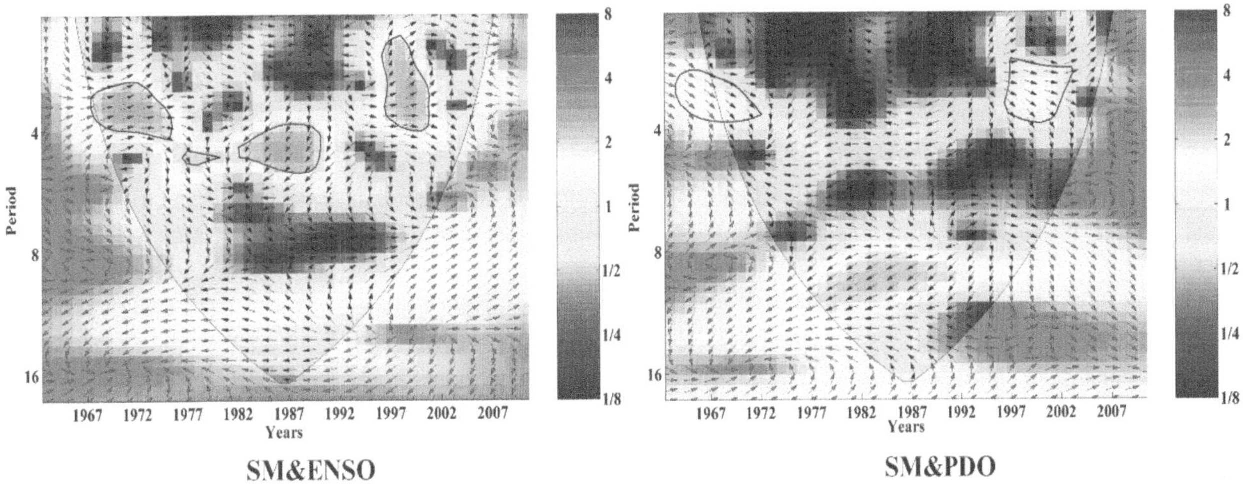

Figure 6 | Cross-wavelet of meteorological drought, underlying surface features and teleconnection factors exceeding the significance level in winter.

correlation of about 2 years in 2004–2007. In winter, the soil moisture and ENSO have a significant positive correlation of about 2–4 years in 1969–1975, a significant negative correlation of about 4–6 years in 1982–1990, a significant positive correlation of about 2–4 years in 1997–2000; the soil moisture and PDO have a significant positive correlation of about 2–4 years in 1967–1972 and 1997–2004.

It can be found from the above results that the teleconnection factors have great influences on meteorological factors and underlying surface features, and then affect the propagation time of drought.

4.4. Quantitative impacts of meteorology, underlying surface features and teleconnection factors on seasonal dynamic propagation

The above section reveals that different seasons have distinct influence factors. For better understanding the relationship between propagation and relevant factors, the MLR is built and the determination coefficient (R^2) is calculated, which are shown in Figure 5. Notably, three scenarios were designed to explore the contribution of meteorological factors, surface features and teleconnection factors. The maximum R^2 of this model fitting is 0.76, which has a certain reliability. The explanation of meteorological factors to the propagation time between meteorological and hydrological drought is 44, 12, 28 and 37% in spring, summer, autumn and winter. Whereas, when the underlying surface features are added, the explanation increased to 49, 27, 37 and 41%, respectively. And when the teleconnection factors were considered, the results have further improved and reached to 57, 54, 76 and 47%. From the above results, it can clearly be seen that the interpretation of propagation time response to underlying surface features is 5, 13, 9 and 4% in different seasons. And the interpretation of propagation time response to teleconnection factors is 8, 27, 39 and 6%, respectively. By comparison, the meteorological factors have the strongest impact on spring and winter, and the teleconnection factors have the strongest impact on summer and autumn. Therefore, the significant slowing drought of propagation patterns in autumn is closely linked with teleconnection factors.

It is obvious that considering the underlying surface features and teleconnection factors will have a better performance of drought propagation, especially in summer and autumn, which is in line with the results outlined in Figures 4 and 6 of section 'Effects of diverse influencing factors on seasonal dynamic propagation', further verifying the reliability of these findings. That is to say, taking the underlying surface features and teleconnection factors into account gives a more accurate interpretation of propagation time response.

5. CONCLUSION

In this study, the propagation dynamics from meteorological to hydrological drought and influence factors have been revealed. The following conclusions can be reached:

- The propagation time from meteorological drought to hydrological drought is generally the quickest in summer (2 months), followed by autumn (3 months), then spring (8 months) and finally winter (13 months).
- The propagation time insignificantly increases in summer and decreases in spring, while significantly increasing in autumn and decreasing in winter.
- The results indicate that AO, ENSO and PDO in autumn; ENSO and PDO in winter have great impacts.

Although the paper has determined the responding time of hydrological drought to meteorological drought in seasons and dynamic changes, and the impacts on propagation process, it also found that considering the teleconnection factors will have a better performance of drought propagation. And the study has provided the basis for drought researchers and forecasters. But the further influencing mechanism has not been revealed. How other factors impact the drought propagation needs to be studied for further investigation.

ACKNOWLEDGEMENTS

This study was jointly funded by the National Key Research and Development Program of China (grant no. 2017YFC0405900), the Strategic Priority Research Program of the Chinese Academy of Sciences (Grant No. XDA28060100), the National Natural Science Foundation of China (grant no. 51709221), the Planning Project of Science and Technology of Water Resources of Shaanxi (grant nos 2015slkj-27 and 2017slkj-19), the Open Research Fund of State Key Laboratory of Simulation and Regulation of Water Cycle in River Basin (China Institute of Water Resources and Hydropower Research, grant no. IWHR-SKL-KF201803), and the Doctorate Innovation Funding of Xi'an University of Technology (grant no. 310-252071712).

DATA AVAILABILITY STATEMENT

All relevant data are available from an online repository or repositories Meteorological data come from China Meteorological Data Network (http://data.cma.cn/). Sunspot data come from International Council for Science World Data System (http://sidc.oma.be/silso/dayssnplot). ENSO (Nino3.4) & AO (Arctic Oscillation) & PDO (Pacific Decadal Oscillation) come from National Oceanic and Atmospheric.

REFERENCES

Allen, R. G., Jensen, M. E., Wright, J. L. & Burman, R. D. 1989 Operational estimates of reference evapotranspiration. *Agronomy Journal* **81** (4), 650–662.

Apurv, T., Sivapalan, M. & Cai, X. 2017 Understanding the role of climate characteristics in drought propagation. *Water Resources Research* **53** (11), 9304–9329.

Brekke, A. 1947 *Drought, Its Causes and Effects, Ivan Ray Tannehill*. Princeton University Press, Princeton, NY. 264 pp. https://doi.org/10.2307/1232917.

Cai, S. Y., Zuo, D. P., Xu, Z. X. & Han, X. M. 2018 Spatiotemporal variability and assessment of drought in the Wei River basin of China. *Proceedings of the International Association of Hydrological Sciences (IAHS)* **379**, 73–82. https://doi.org/10.5194/piahs-379-73-2018.

Chen, H. & Sun, J. 2017 Anthropogenic warming has caused hot droughts more frequently in China. *Journal of Hydrology* **544**, 306–318. https://doi.org/10.1016/j.jhydrol.2016.11.044.

Cook, E. R., Meko, D. M., Stahle, D. W. & Cleaveland, M. K. 1999 Drought reconstruction for the continental United States. *Journal of Climate* **12**, 1145–1162. https://doi.org/10.1175/1520-0442(1999)012,1145:DRFTCU.2.0.CO;2.

Dracup, J. A., Lee, K. S. & Paulson Jr., E. G. 1980 On the definition of droughts. *Water Resources Research* **16** (2), 297–302. https://doi.org/10.1029/wr016i002p00297.

Du, J. W. 2008 *Drought Monitoring Forecasting Assessment and Hazard Management in Shaanxi Province*. China Meteorological Press, Xi'an (in Chinese).

Fang, W., Huang, S. Z., Huang, Q., Huang, G. H., Wang, H., Leng, G. Y., Wang, L. & Guo, Y. 2019a Probabilistic assessment of remote sensing-based terrestrial vegetation vulnerability to drought stress of the Loess Plateau in China. *Remote Sensing of Environment* **232**, 111290. https://doi.org/10.1016/j.rse.2019.111290.

Fang, W., Huang, S., Huang, G., Huang, Q., Wang, H., Wang, L., Zhang, Y., Li, P. & Ma, L. 2019b Copulas-based risk analysis for inter-seasonal combinations of wet and dry conditions under a changing climate. *International Journal of Climatology* **39** (4), 2005–2021. https://doi.org/10.1002/joc.5929.

Feng, P., Hu, R. & Li, J. Z. 2014 Meteorological drought grade prediction using three-dimensional log-linear models. *Journal of Hydraulic Engineering* **45** (5), 505–512.

Fye, F. K., Stahle, D. W. & Cook, E. R. 2010 Paleoclimatic analogs to twentieth-century moisture regimes across the United States. *Bulletin of the American Meteorological Society* **84** (7), 901–910. https://doi.org/10.1175/BAMS-84-7-901.

Ge, Y., Apurv, T. & Cai, X. 2016 Spatial and temporal patterns of drought in the Continental U.S. during the past century. *Geophysical Research Letters* **43** (12), 6294–6303. https://doi.org/10.1002/2016GL069660.

Grinsted, A., Moore, J. C. & Jevrejeva, S. 2004 Application of the cross wavelet transform and wavelet coherence to geophysical time series. *Nonlinear Processes in Geophysics* **11** (5/6), 561–556.

Guo, Y., Huang, S., Huang, Q., Wang, H., Fang, W., Yang, Y. & Wang, L. 2019 Assessing socioeconomic drought based on an improved Multivariate Standardized Reliability and Resilience Index. *Journal of Hydrology* **568**, 904–918. https://doi.org/10.1016/j.jhydrol.2018.11.055.

Huang, S. Z., Chang, J. X., Huang, Q. & Chen, Y. T. 2014a Spatio-temporal changes and frequency analysis of drought in the Wei River Basin, China. *Water Resource Management* **28** (10), 3095–3110. https://doi.org/10.1007/s11269-014-0657-4.

Huang, S. Z., Chang, J. X., Huang, Q. & Chen, Y. T. 2014b Monthly streamflow prediction using modified EMD-based support vector machine. *Journal of Hydrology* **511**, 764–775. https://doi.org/10.1016/j.jhydrol.2014.01.062.

Huang, S. Z., Huang, Q., Wang, Y. M. & Chen, Y. T. 2015 Evolution of drought characteristics in the Weihe River Basin based on standardized precipitation index. *Journal of Natural Disasters* **24**, 15–22.

Huang, S. Z., Li, P., Huang, Q., Leng, G. Y., Hou, B. B. & Ma, L. 2017 The propagation from meteorological to hydrological drought and its potential influence factors. *Journal of Hydrology* **547**, 184–195. https://doi.org/10.1016/j.jhydrol.2017.01.041.

Huang, S., Wang, L., Wang, H., Huang, Q., Leng, G. & Fang, W. 2019 Spatio-temporal characteristics of drought structure across China using an integrated drought index. *Agricultural Water Management* **218**, 182–192. https://doi.org/10.1016/j.agwat.2019.03.053.

Hudgins, L. & Huang, J. 1996 Bivariate wavelet analysis of Asia monsoon and ENSO. *Advances in Atmospheric Sciences* **13** (3), 299–312. https://doi.org/10.1007/bf02656848.

Jiang, C., Wang, F., Xing-Min, M. U. & Rui, L. I. 2012 Temporal and spatial characteristics of climate change and extreme dry and wet events in Wei River Basin in Last 52 years. *Journal of Irrigation and Drainage.* **31** (04), 32–36.

Joo, J., Jeong, S., Zheng, C., Park, C. E., Park, H. & Kim, H. 2020 Emergence of significant soil moisture depletion in the near future. *Environmental Research Letters* **15** (12), 124048 (9 pp).

Leng, G. Y., Tang, Q. H. & Rayburg, S. 2015 Climate change impacts on meteorological, agricultural and hydrological droughts in China. *Global and Planetary Change* **126**, 23–34. https://doi.org/10.1016/j.gloplacha.2015.01.003.

Li, J., Guo, Y., Wang, Y., Lu, S. & Chen, X. 2018 Drought propagation patterns under Naturalized condition using daily hydrometeorological data. *Advances in Meteorology* **2018** (2), 1–14. https://doi.org/10.1155/2018/2469156.

Lorenzo-Lacruz, J., Vicente-Serrano, S. M., Gonzalez-Hidalgo, J. C., Lopez-Moreno, J. I. & Cortesi, N. 2013 Hydrological drought response to meteorological drought in the Iberian Peninsula. *Climate Research* **58** (2), 117–131. https://doi.org/10.3354/cr01177.

McKee, T. B., Doesken, N. J. & Kleist, J. 1993 The Relation of Drought Frequency and Duration to Time Scales. Proceedings of the 8th Conference on Applied Climatology, Anaheim, California, 17–22 January 1993, pp. 179–184.

Mishra, A. K. & Singh, V. P. 2010 A review of drought concepts. *Journal of Hydrology* **391** (1–2), 202–216. https://doi.org/10.1016/j.jhydrol.2010.07.012.

Mohammad, R. K., Majid, V. & Amin, A. 2014 Drought monitoring using a Soil Wetness Deficit Index (SWDI) derived from MODIS satellite data. *Agriculture Water Management* **132** (31), 37–45. https://doi.org/10.1016/10.1016/j.agwat.2013.10.004.

Shukla, S. & Wood, A. W. 2008 Use of a standardized runoff index for characterizing hydrologic drought. *Geophysical Research Letters* **35** (2), 226–236. https://doi.org/10.1029/2007gl032487.

Tallaksen, L. M. & Van Lanen, H. A. J. 2004 Hydrological drought: process and estimation methods for streamflow and groundwater. In: *Developments in Water Science.* 48. Elsevier Science, Amsterdam.

Tallaksen, L. M., Hisdal, H. & Van Lanen, H. A. 2009 Space–time modelling of catchment scale drought characteristics. *Journal of Hydrology* **375** (3), 363–372. https://doi.org/10.1016/j.jhydrol.2009.06.032.

Van Loon, A. F. 2015 Hydrological drought explained. *Wiley Interdisciplinary Reviews: Water* **2** (4), 359–392. https://doi.org/10.1002/wat2.1085.

Van Loon, A. F. & van Lanen, H. A. J. 2012 A process-based typology of hydrological drought. *Hydrology and Earth System Sciences* **16** (7), 1915–1946. https://doi.org/10.5194/hess-16-1915-2012.

Wang, D. B., Mohamad, H., Cai, X. & Valocchi, A. J. 2011 Climate change impact on meteorological, agricultural, and hydrological drought in central Illinois. *Water Resources Research* **47**, 1–13, W09527.

Wang, Y., Li, J., Zhang, T. & Wang, B. 2019 Changes in drought propagation under the regulation of reservoirs and water diversion. *Theoretical and Applied Climatology.* https://doi.org/10.1007/s00704-019-02839-3.

Wilhite, D. A. & Glantz, M. H. 1985 Understanding the drought phenomenon: the rule of definition. *Water International* **10** (3), 111–120. https://doi.org/10.1080/02508068508686328.

World Meteorological Organization (WMO). 1997 *Climate, Drought and Desertification.* WMO No. 869, Geneva, 12 p.

Xu, K., Yang, D., Xu, X. & Lei, H. 2015 Copula based drought frequency analysis considering the spatio-temporal variability in Southwest China. *Journal of Hydrology* **527**, 630–640. https://doi.org/10.1016/j.jhydrol.2015.05.030.

Yu, M., Li, Q., Hayes, M. J., Svoboda, M. D. & Heim, R. R. 2014 Are droughts becoming more frequent or severe in china based on the standardized precipitation evapotranspiration index: 1951–2010. *International Journal of Climatology* **34** (3), 545–558.

Yu, M., Cho, Y., Kim, T. W. & Chae, H. S. 2018 Analysis of drought propagation using hydrometeorological data: from meteorological drought to agricultural drought. *Journal of Korea Water Resources Association* **51** (3), 195–205. https://doi.org/10.3741/JKWRA.2018.51.3.195.

Zhang, B. Q., Wu, P. T., Zhao, X. N., Wang, Y. B., Wang, J. W. & Shi, Y. G. 2012 Drought variation trends in different subregions of the Chinese Loess Plateau over the past four decades. *Agricultural Water Management* **115**, 167–177. https://doi.org/10.1016/j.agwat.2012.09.004.

First received 7 January 2021; accepted in revised form 24 November 2021. Available online 10 December 2021

doi: 10.2166/nh.2021.040

Evaluation of a multisite weather generator on precipitation simulation in the Yangtze river basin

Yu Li[a], Xin-Min Zeng [a,b,*] and Jiali Guo[c,d]

[a] CMA-HHU Joint Laboratory for Hydrometeorological Studies, Hohai University, Nanjing 210098, China
[b] College of Hydrology and Water Resources, Hohai University, Nanjing 210098, China
[c] College of Hydraulic & Environmental Engineering, China Three Gorges University, Yichang 443002, China
[d] Hubei Key Laboratory of Intelligent Yangtze and Hydroelectric Science, China Yangtze Power Co., Ltd, Yi-chang Hubei, 443000, China
*Corresponding author. E-mail: zen_xm@yahoo.com

X-MZ, 0000-0002-7146-0819

ABSTRACT

This paper presents the evaluation of a multisite statistical weather generator (MulGETS: Multisite weather Generator of École de Technologie Supérieure) based on its simulation effect of precipitation in the Yangtze River Basin. MulGETS effectively generates spatially correlated sequences of precipitation simultaneously, while maintaining their spatial and temporal distribution characteristics. On the spatial scales, the accuracy of the model varies from station to station, and in general, the errors are lower at stations in the middle and lower reaches of the Yangtze River Basin than in the upper reaches. This difference is likely to exist because of the lower amount of rainfall and more complex topography than those of the upper river basins. On the temporal scales, the simulated values are more precise on the annual scale than on the seasonal scale. Large relative errors occur more frequently in winter, ranging from −35% to 25%. MulGETS can consistently produce precipitation by considering the intensity, magnitude, and duration indices with sub-basin varied observations. However, the precipitation maxima were much lower than the observations. This work shows the general reasonability of the model in downscaling precipitation in the Yangtze River Basin.

Key words: daily precipitation, MulGETS, spatial distribution, stochastic hydrology, the Yangtze River Basin, weather generator

HIGHLIGHTS

- MulGETS is applied to generate daily precipitation for each subbasin in the Yangtze River Basin.
- The performance is evaluated by 9 indices in three categories over seasonal and annual scales.

INTRODUCTION

The general circulation model (GCM) is currently an important tool for simulating the basin's climate and predicting future climate scenarios. However, the spatial resolution of the model's output data is extremely coarse for application to meteorological and hydrological research at the basin scale. To solve this problem, scientists have extensively researched downscaling methods to convert rough GCM data into high-resolution regional or watershed scale data (Xu 1999; Wilby *et al.* 2002; Wood *et al.* 2004; Mehrotra *et al.* 2014; Pour *et al.* 2014; Singh *et al.* 2017; Araya-Osses *et al.* 2020; Li *et al.* 2020).

In general, downscaling methods can be divided into two categories: dynamical downscaling (Giorgi & Mearns 1999; Zeng *et al.* 2017; Liu *et al.* 2018; Shiferaw *et al.* 2018) and statistical downscaling (Chen *et al.* 2016; Yi *et al.* 2018) methods. Statistical downscaling is widely used because of its flexibility, simplicity, low computational complexity, and the ability to quickly simulate long-term regional climate sequence (Cheng *et al.* 2008; Chen *et al.* 2011; Chen *et al.* 2012; Wang *et al.* 2014; Verdin *et al.* 2015). Owing to the importance of fully understanding the secular variation of hydrological characteristics at the watershed scale, many scholars have applied different statistical downscaling methods to determine hydrological responses to climate change (Ahmed *et al.* 2013; Jeong *et al.* 2013; Wang *et al.* 2013; Lee & Bae 2016; Zheng *et al.* 2018). The climate change information needs to match the hydrological information on the basin scale; in other words, weather data corresponding to the existing stations should be used (Brissette *et al.* 2007).

Weather generators are an important statistical downscaling method, and they are widely used to generate synthetic sequences of weather variables that are statistically consistent with observed variables. Conventional weather generators are based on individual stations (or grid points) to generate single-point data or establish mutually independent regression equations (Brissette *et al.* 2007), however, hydrological simulation results are largely affected by the spatial distribution of regional precipitation. The above-mentioned single-site downscaling methods may neglect the correlation between adjacent points of the forecast elements. This deficiency partly restricts the application of the single-site statistical downscaling method (Chen *et al.* 2014). Therefore, in recent years, many studies have focused on spatially correlated downscaling methods (Kottegoda *et al.* 2003; Raje & Mujumdar 2009; Srikanthan & Pegram 2009; Breinl *et al.* 2013; Jeong *et al.* 2013; Chowdhury *et al.* 2019) or provided useful insights on the development of multisite downscaling methods (Wilks 1998; Harpham & Wilby 2005; Fang & Tacher 2010). Fu evaluated four multisite weather generators to simulate daily precipitation in the Gloucester Watershed in Australia. The results show that the four weather generators can satisfactorily simulate the extreme values of precipitation as well as the duration of wet and dry days (Fu *et al.* 2018).

Chen proposed a MATLAB-based stochastic weather generator referred to as Multisite weather Generator of École de Technologie Supérieure (MulGETS) and tested it by simulating precipitation and temperature at stations in the Jinghe Watershed in China, Peribonka Watershed in Canada, and Walnut Gulch Watershed in America (Chen *et al.* 2014). At these stations, MulGETS was able to retain the spatial correlation of precipitation and temperature between stations. Many researchers have applied MulGETS to areas of interest and compared it with other weather generators. This work has shown that MulGETS often outperformed the other three weather generators when applied to a wathershed of Eastern Ontario, Canada (Alodah & Seidou 2019). Tseng compared MulGETS with three other weather generators on the task of producing precipitation at different spatial scales, and the results showed that MulGETS exhibited the advantage of preserving most of the observed statistics and producing reasonable Q-Q plots (Tseng *et al.* 2020). Vallam and Qin compared MulGETS with a multi-site rainfall simulator, a Neyman–Scott based Poisson cluster model, and a proposed modified *k*-nearest neighbour model for 30 years of rainfall in Singapore, and found that MulGETS exhibited the lowest differences between prediction intervals (Vallam & Qin 2016). MulGETS can also be coupled with other methods to act as a spatial downscaling method. Chen proposed a multi-site downscaling approach coupling a quantile mapping method with MulGETS to simultaneously downscale precipitation for multiple locations within a watershed (Chen *et al.* 2018). He also assessed the effectiveness of MulGETS in hydrological modeling over a Canadian watershed in the province of Québec, and the results showed that the monthly mean discharge was accurately represented by MulGETS-generated precipitation and temperature (Chen *et al.* 2016).

The multisite weather generator effectively solves the challenging problem of retaining the spatial correlation of the input data. However, previous studies on MulGETS have generally concentrated on the consistency between the daily results of model simulations and observations with respect to indices of precipitation intensity and precipitation frequency. Moreover, it is important to evaluate MulGETS results on timescales longer than the daily scale due to the twofold reasons as follows. First, considering possible systematic deviations in some regions, these daily scale biases may be different when calculated on the seasonal or annual scales (e.g., the seasonal biases are likely to be reduced or amplified, or changed from positive to negative), therefore, it is necessary to evaluate MulGETS on the annual and seasonal scales over multiple sub-basins. Second, the Yangtze River Basin is the third-largest river basin in the world, with a total drainage area of approximately 1.8 million km^2, accounting for 18.8% of China's land area, where extraordinary floods or high river water levels of the trunk stream were observed to occur as extreme seasonal hydroclimate events with persistent heavy rainfalls over 1–2 months (e.g., the 1998 and 2020 summer floods in the basin). Meanwhile, the subbasins of the Yangtze River Basin are greatly influenced by frequent floods in which the most severe damage seems to result from specific episodes of extreme precipitation on a regional scale, such as rainstorms and heavy precipitation events that account for high percentages of the yearly total over a few rainy days (Zhang *et al.* 2009). Besides, precise precipitation data, which are sometimes unavailable, are needed for prediction of floods. Hence, in this study, we evaluated the performance of MulGETS using multiple indices of extreme precipitation with MulGETS's ability to reproduce long-term precipitation characteristics examined at daily, annual, and seasonal scales on data collected from the Yangtze River Basin, which we hope will provide a reference for precipitation simulation and flood forecasting in future research.

MATERIALS AND METHODS

Study area

The Yangtze River originates from Tanggula Mountain in Qinghai Province and flows through 11 provinces, administrative regions, or municipalities in China from the west to the east. The river has a total length of 6,380 km and a drainage area of 1.8 million km^2. It is the third-largest drainage basin in the world, and the areas through which it flows contain mountains, plateaus, basins, plains, and other topographies. The entire basin area is subjected to diverse climate types; for example, some areas of the Qinghai-Tibet Plateau climate have a subtropical monsoon climate, the southern part has a tropical climate, and the northern part has a temperate climate. The annual mean precipitation in the Yangtze River Basin is approximately 1,067 mm. However, owing to the vast area and complex topography, the temporal and spatial distribution of annual precipitation and extreme events are very uneven; the annual mean precipitation in the west is between 300 and 500 mm, while that in the southeast can reach 1,600–1,900 mm. In addition, rainfall largely occurs during summertime (Gemmer *et al.* 2008). Based on the climate and underlying surface conditions, we roughly divided the study area into 11 sub-basins from the west to the east. The distribution of the sub-basins and the stations is shown in Figure 1.

The area of each sub-basin and their detailed information are summarized in Table 1, which shows that the stations are evenly distributed. The average density of the stations is approximately 15,000 km^2 per station; however, the density is relatively low at the Jinsha River, Mintuo River, and Jialing River. In addition, three stations in the northern part of the Jialing River were disregarded because of missing essential data, resulting in sparse station distribution in the basin.

Data

The precipitation data of the initial 175 stations in the Yangtze River Basin were obtained from the daily datasets provided by the China Meteorological Data Sharing Service System (http://cdc.cma.gov.cn/index.jsp). After quality control, 123 stations with better continuity and sufficient extent of data were selected, each with complete data from January 1961 to December 2005.

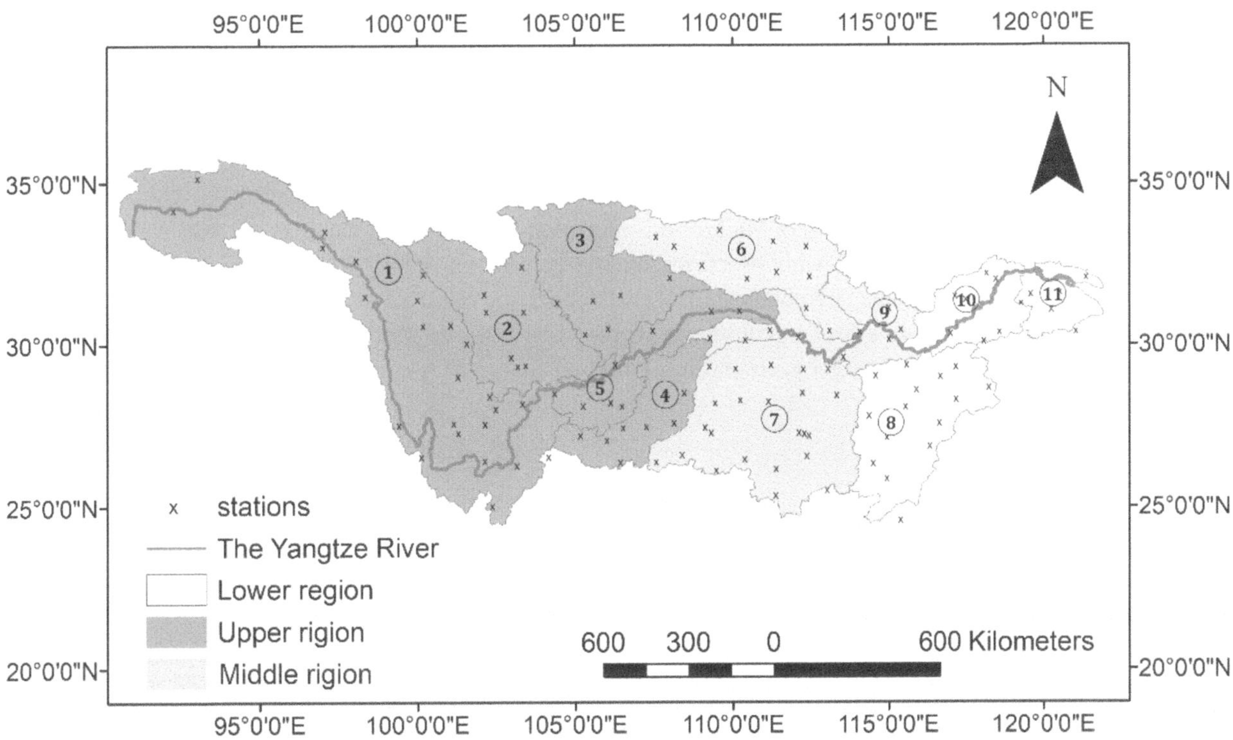

Figure 1 | Distribution of sub-basins and stations of the Yangtze River Basin: (1) Jinshajiang River, (2) Mintuojiang River, (3) Jialingjiang River, (4) Wujiang River, (5) Upper mainstream section, (6) Hanjiang River, (7) Dongting Lake, (8) Poyang Lake, (9) Middle mainstream section, (10) Lower mainstream section, and (11) Tai Lake.

Table 1 | Primary information of sub-basins

Region Categories	Sub-basin	Abbreviation	Area/10^4 km^2	No. of stations	Station density (10^4 km^2/station)
Upper region	Jinsha River	JSR	47.4	20	2.37
	Mintuo River	MTR	16.3	10	1.63
	Jialing River	JLR	16.3	7	2.33
	Wujiang River	WJR	8.9	8	1.11
	Upper mainstream section	UMS	10.3	8	1.29
Middle region	Hanjiang River	HJR	15.8	14	1.23
	Middle mainstream section	MMS	27.2	23	1.18
	Poyang Lake	PYL	17.3	15	1.53
Lower region	Dongting Lake	DTL	10.0	7	1.43
	Lower mainstream section	LMS	9.3	9	1.03
	Tai Lake	THL	4.0	4	1.00
	Total	–	182.8	125	1.46

Model: MulGETS

The weather generator is a random model that generates a comprehensive weather time series with similar statistical characteristics as the input data. When the measured climate data is insufficient to meet the complete spatial coverage, the weather generator can supplement it. The generator can also be used to simulate past or future weather sequences and has become a commonly used tool in the research on climate change and hydrological response (Liao *et al.* 2004; Kilsby *et al.* 2007; Semenov 2008).

MulGETS is a MATLAB-based stochastic weather generator, developed by Chen *et al.* (2014). The MulGETS Matlab routine can be found on the Mathworks file exchange website (https://www.mathworks.com/matlabcentral/fileexchange/47537-multi-site-stochstic-weather-generator-mulgets). The steps and basic principles of MulGETS downscaling for precipitation are as follows:

- Calculate two transition probabilities based on the historical precipitation data of each station.
- The measured precipitation transition probabilities of each station are converted into values in a normal distribution with a mean value of 1 and a standard deviation of 0 by using the inverse complementary error function. Thus, the normalized transition probabilities matrix, norm_trans, is obtained. Then, the Cholesky method is used to generate a spatially correlated random number matrix, corr_random, with a size of m × n (where m is the product of the number of days in a certain month and the number of years in the generated sequence and n is the number of stations in the basin).
- Based on the stability ineffectiveness theory of Markov, assuming that the state of the day is state1, each value of corr_random generated in step 2 is compared with its corresponding parameter, norm_trans. According to the results, the precipitation states of next day, state2, are obtained.
- K_m is introduced in the problem of spatial intermittence proposed by Brissette *et al.* (2007). Then, the daily precipitation occurrence coefficient K_m is calculated for each season as follows:

$$K_m = \frac{O \times C^T}{U \times C^T}$$

  where O is the row vector of indicating whether precipitation occurs in the surrounding stations of a certain station on a certain day, and C^T is the column vector of the seasonal correlation coefficient of precipitation occurrence between a certain station and its surrounding stations. According to the magnitude of K_m, the observed data and its corresponding K_m are divided into 9 intervals, an evenly distributed between 0 and 1, if a class contains fewer than 50 precipitation events, two or more neighboring classes are combined until there are more than 50 events in each of the new classes.
- Parameters for exponential or gamma distribution are then calculated by devising a nonlinear regression equation of the K_m group median and the observed precipitation mean or standard deviation sequence of all points in the interval. Then, the simulated average precipitation is corrected with the observed values, and the parameters of the exponential or gamma distributions are calculated. For gamma distribution, MulGETS calibrates several pairs of (α, β) for each site and season.

- Finally, a random precipitation amount is generated according to the precipitation occurrence coefficient and the related parameters by determining which K_m interval falls on a certain day and generates a certain amount of precipitation based on the interval's parameter and selected distribution.

Experimental design

For each sub-basin, observation data of the stations collected from 1961 to 2005 served as the input data. The MulGETS was used to generate the precipitation series on a daily scale. The only parameters required for the MulGETS operation are the daily precipitation occurrence threshold and type of precipitation distribution function. In this study, 0.1 mm/day is selected as the threshold and the gamma function is the distribution function. The mean and standard deviation of precipitation and occurrence index for each station of each sub-basin are provided in the Supplementary Material.

MulGETS is generally believed to generate a sequence that temporally coincides with the input values. When evaluating MulGETS, all sequences from 1961 to 2005 were used.

Most studies use explanatory variance, root mean squared error, standard deviation, and other indices to evaluate the simulation effect of the model (Huang *et al.* 2010; Liu *et al.* 2011). Some other scholars use more abundant indices, such as quantiles, median, maximum value, and minimum value, for evaluation (Wetterhall *et al.* 2006). In this study, we selected nine indices categorized into three types, which originated from the European Union STARDEX project (STAtistical and Regional dynamical Downscaling of EXtremes for the European region) (Goodess *et al.* 2012). In daily scale evaluation, we reserved standard deviation (SD) and 99th percentile of rain day amounts (Pq99) as extra indices. The abbreviations and definitions of extreme precipitation indices are shown in Table 2.

RESULTS AND DISCUSSION

Overview of simulated daily precipitation

This subsection presents an overview of simulated daily precipitation, in which the SD, 95% percentiles (Pq95), and 99% percentiles (Pq99) were calculated from the MulGETS daily precipitation over 1961–2005.

The spatial distribution of the SD of the observed and simulated precipitation and their relative errors are shown in Figure 2. The SD of simulated precipitation on the daily scale of MulGETS was generally underestimated, and it was 1–5% smaller in most areas, indicating that the dispersion of precipitation series may not be estimated well. However, the situation varied for each sub-basin. At most stations of the middle mainstream and stations near the borders between the Middle mainstream and Wujiang River and the borders between Tai Lake and Lower mainstream, the SD of the simulated values was overestimated by 1–5%, and for several stations was even overestimated by 5–10%. This may be attributed to the steep terrain and mountainous areas in the upper reaches of the Yangtze River, and the heterogeneity of the spatial distribution thereof. In the other sub-basins, there are a few stations with positive RE, which are different from most stations with negative RE. These stations are mostly near the boundary of the basin, which is usually a watershed formed by high mountains. Therefore, it is possible that MulGETS is more likely to produce a more widely distributed precipitation series in mountainous areas and more narrowly distributed ones in plains.

Table 2 | Abbreviations and definitions of the indices

Category	Index	Unit	Definition
	Pint	mm/day	Simple daily intensity (rain per rain day)
Precipitation intensity	Ppdd	%	Mean dry-day persistence
	Pnl90	day	No. of events > long-term 90th percentile
	Pq95	mm	95th percentile of rain day amounts
Precipitation magnitude	Pfl90	mm	Total rainfall from events > long-term 90th percentile
	Px5d	mm	Greatest 5-day total rainfall per year
	Pxcdd	day	Max no. of consecutive dry days
Precipitation persistence	Pn10 mm	day	No. of days with precipitation $\geq$10 mm
	Pdsav	day	Mean dry-day spell lengths

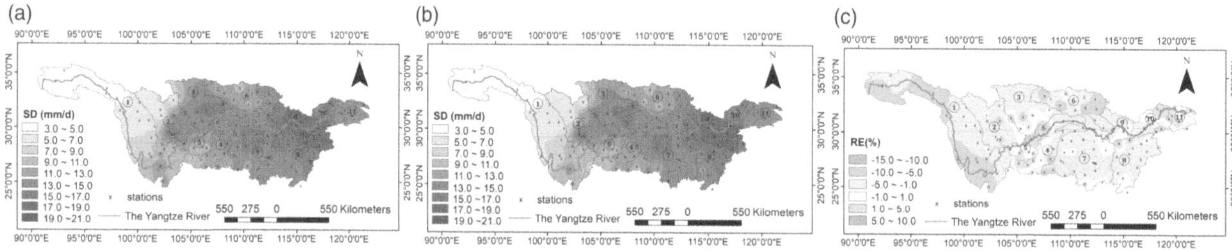

Figure 2 | Spatial distribution of Standard Deviation of (a) the observed data, (b) the MulGETS simulated data, (c) and their relative error of daily precipitation in the Yangtze River Basin (1: Jinshajiang River, 2: Mintuojiang River, 3: Jialingjiang River, 4: Wujiang River, 5: upper mainstream section, 6: Hanjiang River, 7: Dongting Lake, 8: Poyang Lake, 9: middle mainstream section, 10: lower mainstream section, and 11-Tai Lake).

The simulated values of Pq95 and Pq99 were relatively accurate, and the errors in most regions were within 5% (Figure 3). Large negative REs of Pq95 were distributed mainly in the Jinsha River and northern Hanjiang River, while positive REs were located mainly near the borders between the middle mainstream section, Mintuo River, and Jinsha River. Compared with Pq95, Pq99 showed fewer large negative REs, but the spatial distribution of REs was more scattered. In general, in most sub-basins, MulGETS can capture precipitation maximums and reproduce their spatial distribution characteristics.

MulGETS is characterized by its ability to generate precipitation sequences with spatial correlations. The correlation coefficient of precipitation occurrence between stations in the middle and lower reaches of the Yangtze River Basin is larger than that in the upper reaches of the Yangtze River Basin (Figure 4(b1)–4(b11)). MulGETS was able to generate precipitation occurrence sequences consistent with the observed values for all sub-basins, in which all the scattered points were near the 1:1 line. Although with respect to precipitation amounts, the scattered points were farther from the 1:1 line, overall they were still along the line. In the Tai Lake, lower mainstream section, and Poyang Lake areas, the spatial correlation coefficients of simulated values were often lower than that of observation, while those of Jinsha River and Wujiang River were higher than observation. On a daily scale, a good simulation of precipitation occurrence cannot guarantee a good spatial correlation of precipitation amounts. The interstation correlation is induced at the monthly scale for precipitation occurrence and at the seasonal scale for precipitation amounts, so the accuracy of spatial correlation on the daily scale may not be very high. When evaluating the effect of the weather generator, the indices calculated on different time scales may exhibit different performance characteristics, which are determined by the construction and properties of the model.

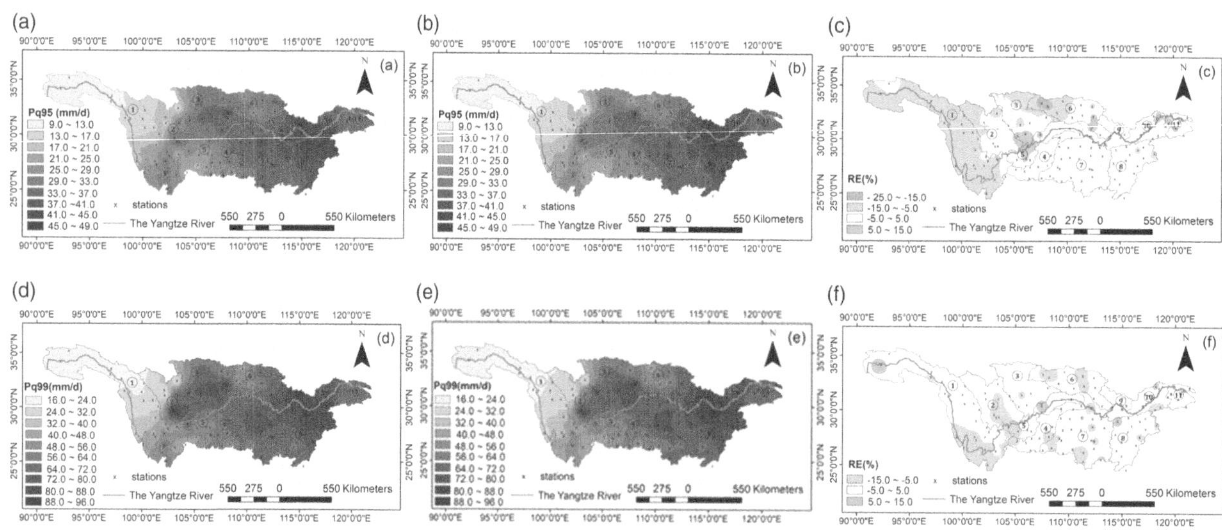

Figure 3 | Spatial distribution of Pq95&Pq99 of (a) & (d) the observed data, (b) & (e) the MulGETS simulated data, (c) & (f) and their relative error of daily precipitation in the Yangtze River Basin (1: Jinshajiang River, 2: Mintuojiang River, 3: Jialingjiang River, 4: Wujiang River, 5: upper mainstream section, 6: Hanjiang River, 7: Dongting Lake, 8: Poyang Lake, 9: middle mainstream section, 10: lower mainstream section, and 11-Tai Lake).

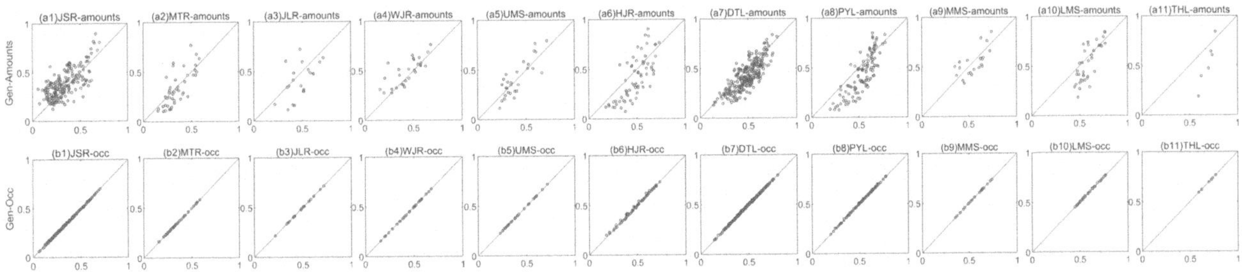

Figure 4 | Observed correlation of precipitation amounts vs. Generated correlation (top) and Observed correlation of precipitation occurrence vs. Generated correlation occurrence (bottom) for all station pairs over all 11 watersheds.

Annual-scale precipitation

In the 11 sub-basins of the Yangtze River Basin, MulGETS was used to downscale precipitation from 1961 to 2005. The simulation results were evaluated using 9 indices and analysed at both annual and seasonal scales. At the seasonal scale, summer and winter were selected as representatives.

Spatial distribution of pint

The Pint of the Yangtze River Basin has an increasing pattern from the west to the east (Figure 5(a)). The observed precipitation amount in the lower reaches is greater than that in the middle reaches, and it is higher in the middle reaches than in the upper reaches. Moreover, stations on the south and the north sides of the basin were rainier. The comparison of Figure 5(a) and 5(b) shows that MulGETS can reasonably simulate the Pint of precipitation at most stations in the Yangtze River Basin and retain the increasing pattern of low and high values in the west and east, respectively. The spatial distribution of the simulated values was also largely consistent with the observed values. The simulation effects of each sub-basin were different.

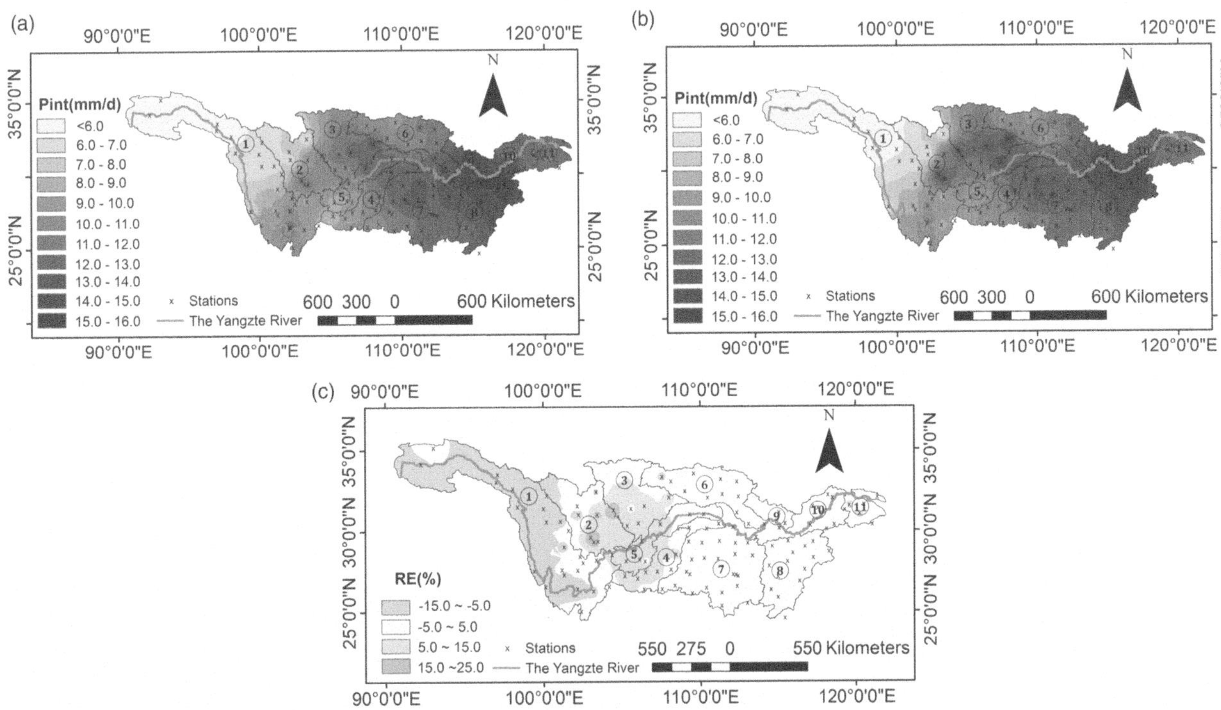

Figure 5 | Spatial distribution of Pint of (a) the observed data, (b) the MulGETS simulated data, (c) and their relative error of annual precipitation in the Yangtze River Basin (1: Jinshajiang River, 2: Mintuojiang River, 3: Jialingjiang River, 4: Wujiang River, 5: upper mainstream section, 6: Hanjiang River, 7: Dongting Lake, 8: Poyang Lake, 9: middle mainstream section, 10: lower mainstream section, and 11-Tai Lake).

As shown in Figure 5(c), the simulation errors of most stations were between −5 and 5%. The stations with errors below −5% were mainly located in the Jinsha River. In comparison, the stations with errors more significant than 5% were mainly located in the Chongqing and Sichuan area, eastern Mintuo River, southern Jialing River, western Wujiang River, and western upper mainstream. This region is the junction of mountains and plains; therefore, precipitation at different stations in the same sub-basin varies greatly, leading to precipitation homogenization at some stations and sizeable relative errors of simulation values.

Performance of various indices

The boxplots of each index are shown in Figure 6(a)–6(i). Overall, MulGETS has a good simulation effect on each index, and the box line of the simulated value is generally consistent with the observations.

MulGETS can simulate the precipitation intensity indices accurately on an annual scale (Figure 6(a)–6(c)). It can also reasonably reproduce each station's Pint characteristics (Figure 6(a)). For example, the observed and simulated values of each station are in good agreement, and the characteristics of different stations (even adjacent ones) were also well distinguished. In most sub-basins, Ppdd shows a slight difference between stations; MulGETS can accurately capture the differences between stations in other basins, such as the Ppdd in some stations in the Mintuo and Jinsha River. However, at stations where Ppdd has high interannual variation, the simulated Ppdd shows lower variability; this indicates that the error will be larger in years when drought time is overconcentrated or scattered. In general, Ppdd in these basins is above 50%, and most of them reach 80%, indicating a high proportion of consecutive dry days. Pnl90 reflects the total number of precipitation days. As shown, Pnl90 has an inverse correlation with Ppdd. MulGETS slightly underestimated the interannual variance in Pnl90.

Clearly, the precipitation magnitude indices of the simulated values are mostly consistent with the observations; however, the range of boxplots at some stations is slightly different from the actual values (Figure 6(d)–6(f)). Specifically, there was a good correspondence between Pq95 and Pint. As shown in Figure 6(a) and 6(d), stations with high Pint are more likely to have a higher Pq95, and the interannual variation of Pq95 is greater than that of Pint. MulGETS also have good simulating effect on Pq95 (Figure 8(d)). For stations where precipitation varies widely from year to year, the simulated values show the same features and vice versa. However, the simulation of Pnl90 is much larger overall, especially in the Jinsha River, Mintuo River, Jialing River, Wujiang River, Hanjiang River, and Dongting Lake, most of which are in the middle and upper reaches of

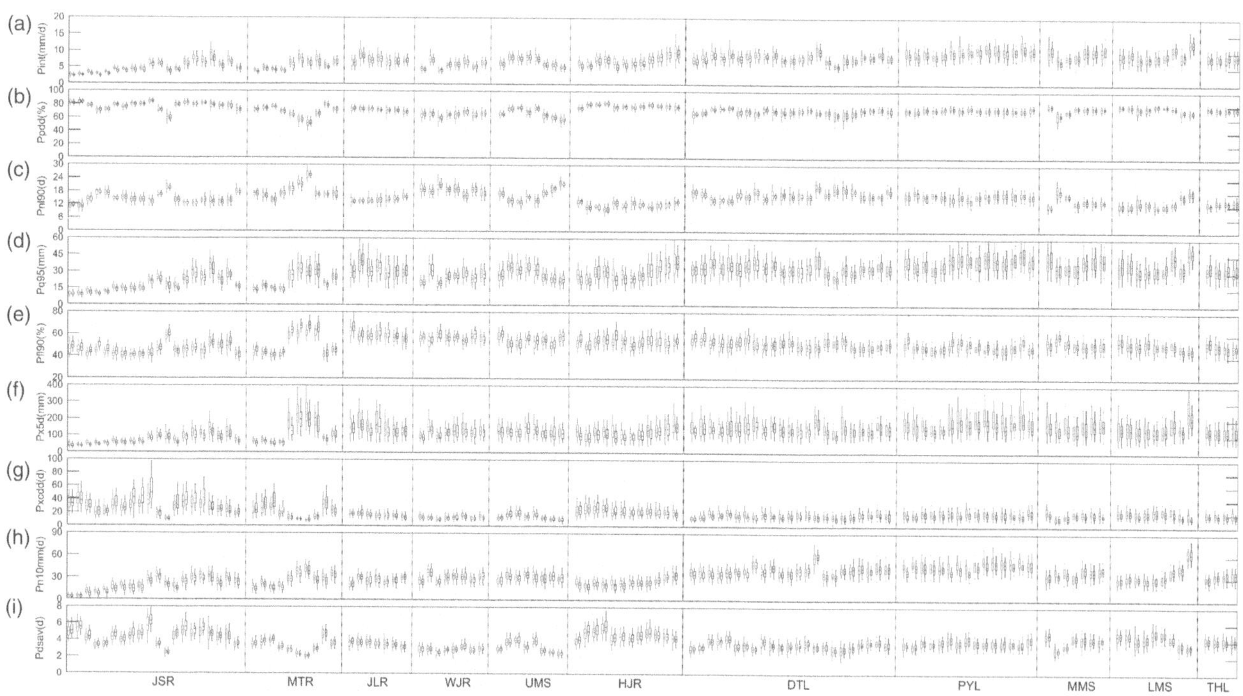

Figure 6 | Comparison between indices of annual scale observation and MulGETS simulated precipitation.

the Yangtze River Basin. The errors of Pnl90 seem to indicate that overestimating the 10% largest part of precipitation is one of the reasons for the systematic errors. However, Px5d was underestimated in most basins. These two inverse errors may cancel each other to ensure a more accurate overall mean, such as Pint.

Most of the annual scale precipitation persistence indices are well simulated; however, in some basins, a systematic error was observed (Figure 6(g)–6(i)). Stations with a relatively medium Pxcdd had a more accurate Pxcdd (Figure 6(g)–6(i)). For example, the simulated Pxcdd values for some stations in the Jinsha River and Mintuo River are slightly overestimated; for the Jialing River, Hanjiang River, the lower mainstream, and Tai Lake they are accurate; for Wujiang River, Dongting Lake, and the middle mainstream basins, they are all low. Because Pxcdd is underestimated in the sub-basins with lower observed Pxcdd, the influence of the simulated errors is comparably small. However, for the stations where Pxcdd is high, overestimation will result in more drought events. Pn10 mm is, to some extent, consistent with Pq95, which has a good relationship with Pint. MulGETS's simulation value for Pn10 mm is accurate at most stations and slightly low at some of the Jinsha River, Mintuo River, and Hanjiang River stations. Pdsav is also well simulated at most stations, and MulGETS precisely captures its interannual variation and interstation differences. Moreover, Pdsav has a positive relationship with Ppdd (Figure 6(b)). However, the Ppdd simulated values are not as precise as that of Pdsav for MulGETS, thus underestimating its interannual variation.

Seasonal-scale precipitation

Spatial distribution of pint

On a seasonal scale, the simulated Pint is in good agreement with the observed ones, and the differences between each sub-basin and the overall distribution can be well simulated. In the Yangtze River Basin, summer precipitation accounted for more than half of the annual precipitation and the Pint is approximately 7.0–24.5 mm/day; in winter, it is less than 10 mm/day in general. The interannual mean precipitation has a spatial distribution of gradually increasing from the upper reaches to the lower reaches, especially in the winter (Figure 7(d)).

In summer, the Sichuan area receives a considerable amount of precipitation; at some stations, the amount can be equivalent to the lower reaches (Figure 7(a)). As shown in Figure 7(a) and 7(b), the precipitation conditions of some stations differ from those of the surrounding stations, forming an evident local closure; for example, some stations in the Jinsha River and Hanjiang River. MulGETS can capture such differences (Figure 7(c) and 7(d)), indicating that the model does not provide a completely average precipitation.

At most stations, the relative error of Pint was between −5 and 5%, and the simulation effect was good (Figure 7(e) and 7(f)). The negative values were mainly concentrated in the upstream basins. In summer, the simulated Pint is overestimated in most sub-basins, whereas in winter, only a few stations have overestimated simulation values in the Jialing River, Hanjiang River, and Wujiang River. Unlike the annual scale simulation, the positive relative errors in summer and winter are scattered rather than concentrated in the Sichuan area. Because there is less precipitation in winter, the denominator of the relative error calculation is smaller. If the absolute error is the same, the relative error will be much larger than that in summer. Therefore, the distribution area of stations with significant errors in winter was larger than that in summer throughout the year (Figure 8(f)).

MulGETS can closely simulate the spatial distribution of Pint in the Yangtze River Basin; however, the relative error at some stations is still difficult to ignore. The accuracy of the model should be increased, especially in stations with less precipitation.

Performance of various indices

The boxplots of the seasonal scale indices are shown in Figures 8 and 9. It was found that MulGETS can precisely simulate precipitation at a seasonal scale. For the precipitation intensity indices, the overall effect is good, especially for Ppdd and Pnl90, which are highly consistent with the observations. In the summer, the Pint is larger than the annual mean; in the winter, it is rather small and mostly below 10 mm/day. MulGETS has a good simulation effect in both seasons (Figures 8(a) and 9(a)). Nevertheless, it faces certain problems, in several sub-basins, the interannual variation of simulated values at various stations tends to be consistent. For the Jialing River in Figure 8(a), the maximum interannual observations were not captured. In winter, the Jinsha River as well as the Mintuo River, where the observations are low, have inaccurate Pint's simulated values; the difference between the stations is not captured as well. The simulated summer Ppdd was lower than that of the annual scale, but its interannual variation range was higher than that of the annual scale (Figure 8(b)), which agrees with

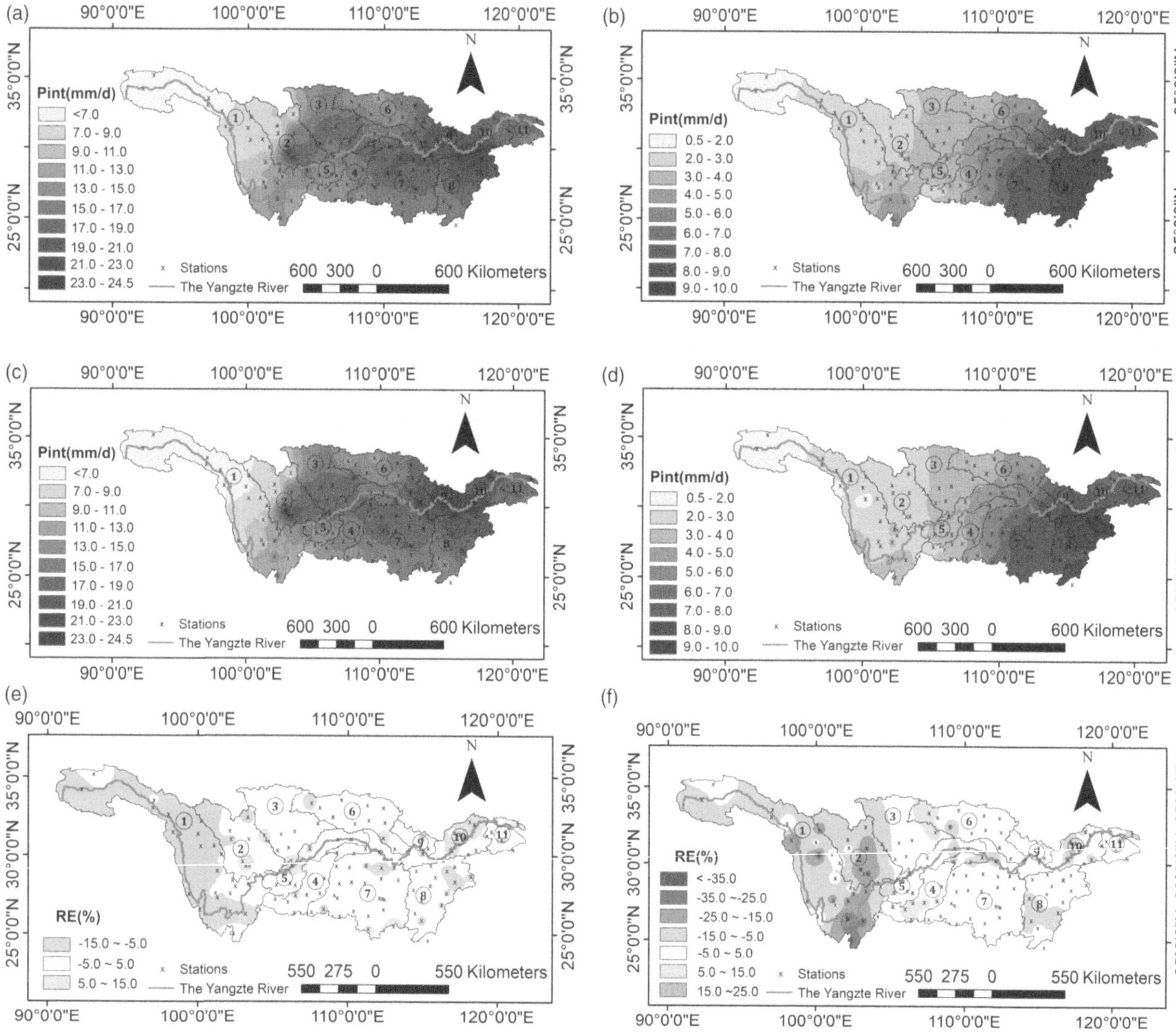

Figure 7 | Spatial distribution of Pint of observed data (a, d), MulGETS simulated data (b, e), and their relative error (c, f) in the summer (the left column) and the winter (the right column) precipitation in the Yangtze River Basin (1: Jinshajiang River, 2: Mintuojiang River, 3: Jialingjiang River, 4: Wujiang River, 5: upper mainstream section, 6: Hanjiang River, 7: Dongting Lake, 8: Poyang Lake, 9: middle mainstream section, 10: lower mainstream section, and 11: Tai Lake).

the observations. This is probably because summer precipitation in the Yangtze River Basin is affected by large-scale ocean circulation and sea surface temperature anomalies such as Hardley circulation and ENSO (El Niño-Southern Oscillation) (Zhou & Wang 2006; Luo *et al.* 2018), which lead to large interannual fluctuations during the summer precipitation. The simulated Pnl90 results are very consistent with the observation (Figure 8(c)); this is probably because the model refers to the observed rainy/no-rain days and only activates rainfall simulation on rainy days. Therefore, with a steady total number of days, the number of rain days will be close to the observation, as will the Pnl90. The shorter the period, the more concentrated is Pnl90, and thus the boxplot is compressed into a single line (even during winter). This index shows interannual differences on an annual scale (with a longer time range), although its interannual variation is underestimated. In winter, MulGETS can reproduce Ppdd and Pnl90 in most basins (Figure 9(b) and 9(c)). Overall, there is a need to improve the simulation effect of MulGETS for winter precipitation.

The range of seasonal scale precipitation magnitude indices is wide, but MulGETS can simulate each sub-basin's characteristics and the differences between stations. Compared with the annual scale Pq95, the simulated Pq95 in the summer was not significantly lower. However, in some sub-basins, the interannual variation in simulated values is underestimated. For example, in the Jialing River (Figure 8(d)), the interannual maximum values of some stations were not captured accurately.

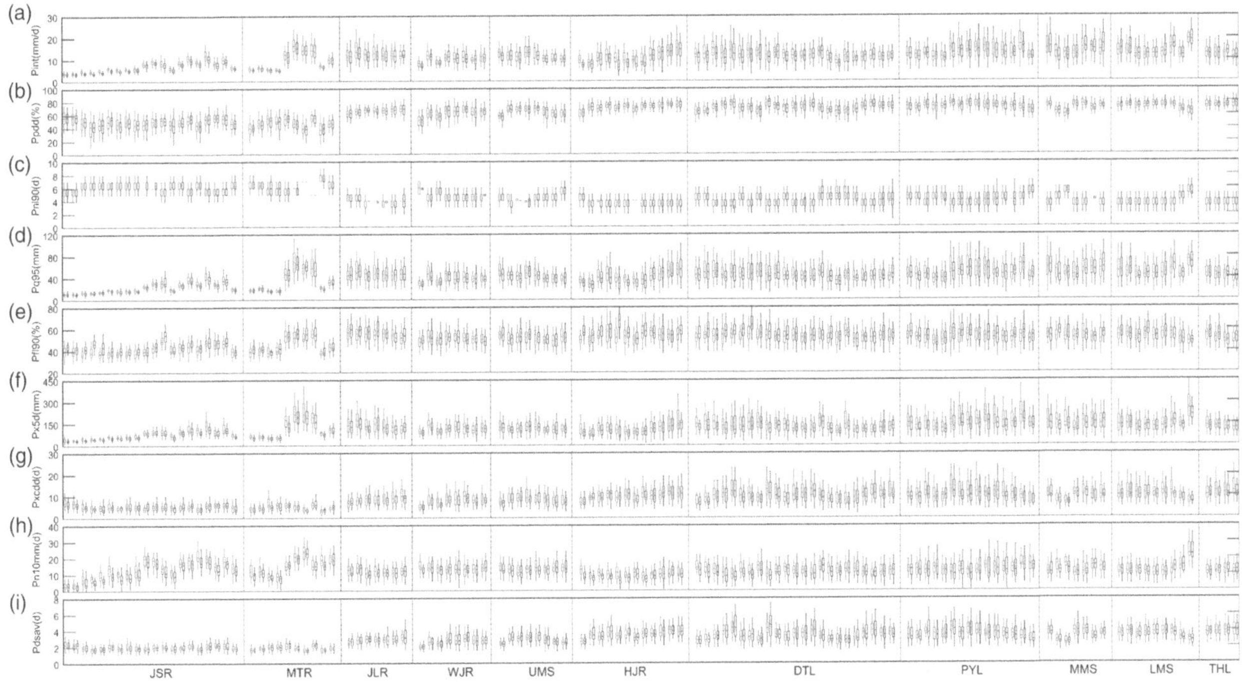

Figure 8 | Comparison between indices of observation and MulGETS simulated precipitation in summer.

The simulation results of Pq95 in the winter are in good agreement with those of Pint, and there is a significant error in some of the Jinsha River and Mintuo River stations. For a basin with less precipitation in the winter, most rain days may be concentrated in a few short periods so that the error of Pq95 will lead to a corresponding error of Pint. In the summer, the Pfl90 is slightly overestimated in each sub-basin, indicating that MulGETS simulation values have a systematic error in this index (Figures 8(e) and 9(e)). Pfl90 is larger than Pnl90, but Pnl90 is more accurate, indicating that the simulation value of MulGETS is not accurate enough to capture the distribution of daily precipitation in the summer. In summer, Px5d was generally underestimated in each sub-basin (Figure 8(f)). In winter, the effect was not good. Specifically, the Px5d and its interannual variation are underestimated in the Jinsha River, Mintuo River, Jialing River, Wujiang River, and the mainstream basins of the upper reaches of the Yangtze River. In summary, on a seasonal scale, MulGETS can simulate the differences between stations but is still weak in heavy rainfall simulations. This is a disadvantage when applying the model to precipitation prediction and flood risk management in the Yangtze River Basin.

The simulation effects were varied for the precipitation persistence indices. In general, the simulation effect of Pxcdd in the middle and lower reaches of the Yangtze River Basin is better than that in the upper reaches of the Jinsha River, Jialing River, Mintuo River, and Wujiang River (Figure 8(g)). For example, at the Jialing River Basin's first station, the simulated value was approximately 50% higher than the maximum of inputs and, at the seventh station, 50% lower than the maximum value. Pxcdd was underestimated in most basins in winter; however, the difference was not significant (Figure 9(g)). Pn10 mm is generally underestimated in the Jinsha and Mintuo River and partly underestimated in the Hanjiang River and Dongting Lake. The minimum interannual values were all lower than the observation minima (Figure 8(h)). Some simulated values reached zero in the middle and upper reaches in the winter. Furthermore, when the observation is zero, the simulated result must be zero but not vice versa (Figure 9(h) and 9(i)). This inclination indicates that although the MulGETS simulation result at one station is affected by its surrounding stations, its own precipitation condition still plays a decisive role. The Pdsav of the simulated value is in good agreement with the observation. However, the interannual variance is overestimated at some stations, such as stations in the Jialing River, Hanjiang River, and Dongting Lake (Figure 8(i)). In summary, MulGETS tends to underestimate the number of days with precipitation of more than 10 mm in the summer and overestimate the most extended length without rain, which is also one reason for some other indices' errors.

In summary, owing to the underestimation of heavy rain in summer and the overestimation of rainless days, MulGETS tends to produce a precipitation series that is drier than the observation. However, the extent is not the same in the sub-basins.

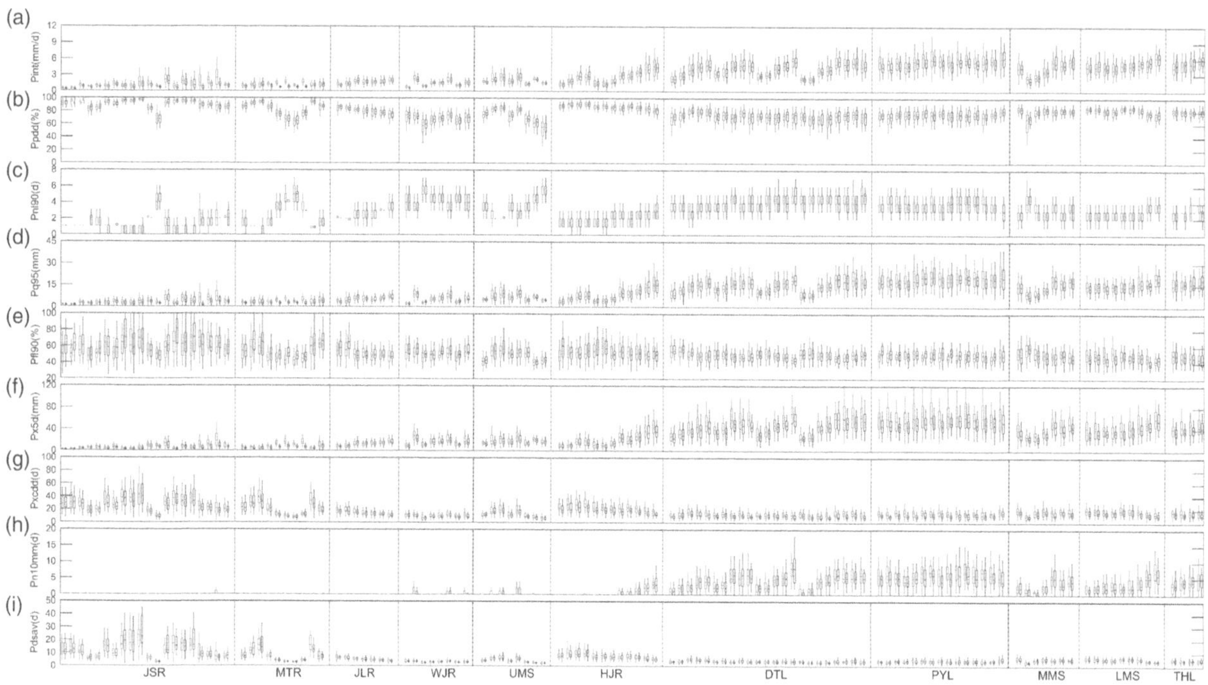

Figure 9 | Comparison between indices of observation and MulGETS simulated precipitation in winter.

Specifically, the simulation effect at the stations at the middle and lower reaches, with more precipitation, is better than that at the upper reaches with less precipitation.

CONCLUSIONS

The primary purpose of the weather generators is to reconstruct data with statistical features close to the input observation data, where one crucial and challenging factor is to reconstruct the time autocorrelation and space correlation of the data. We evaluated MulGETS, a multisite stochastic weather generator, for its simulation effect on precipitation in the Yangtze River Basin from the perspective of both time and space. The results showed that MulGETS can effectively generate spatially correlated variables. The simulated sequences retained the spatial distribution and temporal variation of the input data to a great extent. The simulation effect of the model varied for each sub-basin. The simulation effect of the model for the basins at the middle and lower reaches was better than that for the upper reaches. A possible explanation is the scanty rainfall conditions and the influence of mountainous topography in the upper reaches. On the time scale, the model performs better on the annual scale than on the seasonal scale, and it performs better in summer than in winter.

MulGETS provides satisfactory results for the simulation of intensity, magnitude, and duration indices and is more accurate at simulating the indices reflecting rainless days. However, its ability to capture maximum values must be improved.

In addition, we proposed a method to evaluate multisite downscaling models by analysing the effects of simulations in terms of both spatial distribution and temporal variation. This study provides a reference not only to apply MulGETS in the Yangtze River Basin but also to understand and improve the MulGETS. This study only examined the simulation effect of the MulGETS model. The next step will be to add results from other models for comparison, which will help to further understand the characteristics of the multisite downscaling model.

ACKNOWLEDGEMENTS

We gratefully acknowledge the source code of the MulGETS shared by Professor Chen. The authors would like to thank two anonymous reviewers for their helpful comments and suggestions.

AUTHOR CONTRIBUTIONS

Conceptualization, Yu Li., X. M. Zeng, J.L. Guo; methodology, Yu Li, X. M. Zeng, J. L. Guo; formal analysis, Yu Li; writing – original draft preparation, Yu Li; writing – review and editing, Yu Li, X. M. Zeng; visualization, Yu Li; supervision, X. M. Zeng; funding acquisition, X. M. Zeng. All authors have read and agreed to the published version of the manuscript.

FUNDINGS

This research was funded by the National Key Research and Development Program of China, grant number 2018YFC1508200, the National Natural Science Foundation of China, grant number 42005023, the Fundamental Research Funds for the Central Universities, grant number B210203067 and Open Funding of Hubei Key Laboratory of Intelligent Yangtze and Hydroelectric Science, China Yangtze Power Co.,Ltd,(NO. ZH2002000103).

COMPETING INTEREST

The authors declare that they have no known competing financial interests or personal relationships that could have appeared to influence the work reported in this paper.

DATA AVAILABILITY STATEMENT

All relevant data are available from an online repository or repositories (http://data.cma.cn/).

REFERENCES

Ahmed, K. F., Wang, G. L., Silander, J., Wilson, A. M., Allen, J. M., Horton, R. & Anyah, R. 2013 Statistical downscaling and bias correction of climate model outputs for climate change impact assessment in the U.S. northeast. *Global and Planetary Change* **100** (2013), 320–332.

Alodah, A. & Seidou, O. 2019 The adequacy of stochastically generated climate time series for water resources systems risk and performance assessment. *Stochastic Environmental Research and Risk Assessment* **33** (1), 1–17.

Araya-Osses, D., Casanueva, A., Román-Figueroa, C., Uribe, J. M. & Paneque, M. 2020 Climate change projections of temperature and precipitation in Chile based on statistical downscaling. *Climate Dynamics* **54** (9–10), 4309–4330.

Breinl, K., Turkington, T. & Stowasser, M. 2013 Stochastic generation of multi-site daily precipitation for applications in risk management. *Journal of Hydrology* **498** (2013), 23–35.

Brissette, F. P., Khalili, M. & Leconte, R. 2007 Efficient stochastic generation of multi-site synthetic precipitation data. *Journal of Hydrology* **345** (3–4), 121–133.

Chen, J., Brissette, F. P. & Leconte, R. 2011 Assessment and improvement of stochastic weather generators in simulating maximum and minimum temperatures. *Transactions of the ASABE* **54** (5), 1627–1637.

Chen, H., Xu, C. Y. & Guo, S. L. 2012 Comparison and evaluation of multiple GCMs, statistical downscaling and hydrological models in the study of climate change impacts on runoff. *Journal of Hydrology* **434–435**, 36–45.

Chen, J., Brissette, F. & Zhang, X. J. 2014 A multi-site stochastic weather generator for daily precipitation and temperature. *Transactions of the ASABE* **57** (5), 1375–1391.

Chen, J., Brissette Francois, P. & Zhang, X. C. 2016 Hydrological modeling using a multisite stochastic weather generator. *Journal of Hydrologic Engineering* **21** (2), 04015060.

Chen, J., Chen, H. & Guo, S. L. 2018 Multi-site precipitation downscaling using a stochastic weather generator. *Climate Dynamics* **50** (5), 1975–1992.

Cheng, C. S., Li, G., Li, Q. & Auld, H. 2008 Statistical downscaling of hourly and daily climate scenarios for various meteorological variables in South-central Canada. *Theoretical & Applied Climatology* **91** (1–4), 129–147.

Chowdhury, A. K., Kar, K. K., Shahid, S., Chowdhury, R. & Rashid, M. M. 2019 Evaluation of spatio-temporal rainfall variability and performance of a stochastic rainfall model in Bangladesh. *International Journal of Climatology* **39** (11), 4256–4273.

Fang, J. N. & Tacher, L. 2010 An effcient and accurate algorithm for generating spatially-correlated random fields. *Communications in Numerical Methods in Engineering* **19** (10), 801–808.

Fu, G. B., Chiew, F. H. S. & Shi, X. G. 2018 Generation of multi-site stochastic daily rainfall with four weather generators: a case study of Gloucester catchment in Australia. *Theoretical & Applied Climatology* **134** (2018), 1027–1046.

Gemmer, M., Jiang, T., Su, B. D. & Kundzewicz, Z. W. 2008 Seasonal precipitation changes in the wet season and their influence on flood/drought hazards in the Yangtze River Basin. *China Quaternary International* **186** (No.1), 12–21.

Giorgi, F. & Mearns, L. O. 1999 Introduction to special section: regional climate modeling revisited. *Journal of Geophysical Research* **104** (D6), 6335–6352.

Goodess, C., Anagnostopoulou, C., Bárdossy, A., Frei, C., Harpham, C., Haylock, M., Hundecha, Y., Maheras, P., Ribalaygua, J., Schmidli, J., Schmith, T., Tolika, K. & Tomozeiu, R. 2012 An intercomparison of statistical downscaling methods for Europe and European regions – assessing their performance with respect to extreme temperature and precipitation events. *Climatic Change*.

Harpham, C. & Wilby, R. L. 2005 Multi-site downscaling of heavy daily precipitation occurrence and amounts. *Journal of Hydrology* **312** (1), 235–255.

Huang, J., Zhang, J. C., Zhang, Z. X., Xu, C. Y., Wang, B. L. & Yao, J. 2010 Estimation of future precipitation change in the Yangtze river basin by using statistical downscaling method. *Stochastic Environmental Research and Risk Assessment* **25** (6), 781–792.

Jeong, D. I., St-Hilaire, A., Ouarda, T. & Gachon, P. 2013 Projection of future daily precipitation series and extreme events by using a multi-site statistical downscaling model over the great Montréal area, Québec, Canada, Canada. *Hydrology Research* **44** (1), 147–168.

Kilsby, C. G., Jones, P. D., Burton, A., Ford, A. C., Fowler, H. J., Harpham, C., James, P., Smith, A. & Wilby, R. L. 2007 A daily weather generator for use in climate change studies. *Environmental Modelling & Software* **22** (12), 1705–1719.

Kottegoda, N. T., Natale, L. & Raiteri, E. 2003 A parsimonious approach to stochastic multisite modelling and disaggregation of daily rainfall. *Journal of Hydrology* **274** (1–4), 47–61.

Lee, M.-H. & Bae, D.-H. 2016 Uncertainty assessment of future high and low flow projections according to climate downscaling and hydrological models. *Procedia Engineering* **154** (2016), 617–623.

Li, J. F., Gan, T. Y., Chen, Y. D., Gu, X., Hu, Z. Y., Zhou, Q. M. & Lai, Y. C. 2020 Tackling resolution mismatch of precipitation extremes from gridded GCMs and site-scale observations: implication to assessment and future projection. *Atmospheric Research* **293**, 104908.

Liao, Y., Zhang, Q. & Chen, D. 2004 Precipitation simulation in China with a weather generator. *Acta Geographica Sinica* **59** (5), 689–698.

Liu, Z. F., Xu, Z. X., Charles, S. P., Fu, G. B. & Liu, L. 2011 Evaluation of two statistical downscaling models for daily precipitation over an arid basin in China. *International Journal of Climatology* **31** (13), 2006–2020.

Liu, P., Qiu, X. B., Yang, Y., Ma, Y. Y. & Jin, S. L. 2018 Assessment of the performance of three dynamical climate downscaling methods using different land surface information over China. *Atmosphere* **9** (16), 101.

Luo, J. F., Chen, M., Li, Y., Peng, T. & Xu, H. T. 2018 Influence of the two types of ENSO events on seasonal precipitation over middle and lower reaches of the Yangtze River. *South-to-North Water Trasfers and Water Science&Technology* **016** (004), 82–89.

Mehrotra, R., Evans, J. P., Sharma, A. & Sivakumar, B. 2014 Evaluation of downscaled daily rainfall hindcasts over Sydney, Australia using statistical and dynamical downscaling approaches. *Hydrology Research* **45** (2), 226–249.

Pour, S. H., Harun, S. B. & Shahid, S. 2014 Genetic programming for the downscaling of extreme rainfall events on the east coast of peninsular Malaysia. *Atmosphere* **5** (4), 914–936.

Raje, D. & Mujumdar, P. P. 2009 A conditional random field-based downscaling method for assessment of climate change impact on multisite daily precipitation in the Mahanadi basin. *Water Resources Research* **45** (10).

Semenov, M. A. 2008 Simulation of extreme weather events by a stochastic weather generator. *Climate Research* **35** (3), 203–212.

Shiferaw, A., Tadesse, T., Rowe, C. & Oglesby, R. 2018 Precipitation extremes in dynamically downscaled climate scenarios over the Greater Horn of Africa. *Atmosphere* **9** (3), 112.

Singh, V., Sharma, A. & Goyal, M. K. 2017 Projection of hydro-climatological changes over eastern himalayan catchment by the evaluation of RegCM4 RCM and CMIP5 GCM models. *Hydrology Research* **50** (1), 117–137.

Srikanthan, R. & Pegram, G. G. S. 2009 A nested multisite daily rainfall stochastic generation model. *Journal of Hydrology* **371** (1), 142–153.

Tseng, S. C., Chen, C. J. & Senarath, S. 2020 Evaluation of multi-site precipitation generators across scales. *International Journal of Climatology* **40** (10), 4622–4637.

Vallam, P. & Qin, X. S. 2016 Multi-site rainfall simulation at tropical regions: a comparison of three types of generators. *Meteorological Applications* **23** (3), 425–437.

Verdin, A., Rajagopalan, B., Kleiber, W., Podestá, G. & Bert, F. 2015 A conditional stochastic weather generator for seasonal to multi-decadal simulations. *Journal of Hydrology* **556** (2018), 835–846.

Wang, W. G., Xing, W. Q., Shao, Q. X., Yu, Z. B., Peng, S. Z., Yang, T., Yong, B., Taylor, J. & Singh, V. P. 2013 Changes in reference evapotranspiration across the Tibetan Plateau: observations and future projections based on statistical downscaling. *Journal of Geophysical Research: Atmospheres* **118** (10), 4049–4068.

Wang, W. G., Yu, Z. B., Zhang, W., Shao, Q. X., Zhang, Y. W., Luo, Y. F., Jiao, X. Y. & Xu, J. Z. 2014 Responses of rice yield, irrigation water requirement and water use efficiency to climate change in China: historical simulation and future projections. *Agricultural Water Management* **146** (2014), 249–261.

Wetterhall, F., Bárdossy, A., Chen, D., Halldin, S. & Xu, C. Y. 2006 Daily precipitation-downscaling techniques in three Chinese regions. *Water Resources Research* **42** (11), W11423.

Wilby, R. L., Dawson, C. W. & Barrow, E. M. 2002 SDSM – a decision support tool for the assessment of regional climate change impacts. *Environmental Modelling & Software* **17** (2), 145–157.

Wilks, D. S. 1998 Multisite generalization of a daily stochastic precipitation generation model. *Journal of Hydrology* **210** (1–4), 178–191.

Wood, A. W., Leung, L. R., Sridhar, V. & Lettenmaier, D. P. 2004 Hydrologic implications of dynamical and statistical approaches to downscaling climate model outputs. *Climatic Change* **62** (1–3), 189–216.

Xu, C. Y. 1999 From GCMs to river flow: a review of downscaling methods and hydrologic modelling approaches. *Progress in Physical Geography: Earth and Environment* **23** (2), 229–249.

Yi, C., Shin, Y. & Roh, J.-W. 2018 Development of an urban high-resolution Air temperature forecast system for local weather information services based on statistical downscaling. *Atmosphere* **9** (5), 164.

Zeng, X. M., Wang, M., Wang, N., Yi, X., Chen, C. H., Zhou, Z. G., Wang, G. L. & Zheng, Y. Q. 2017 Assessing simulated summer 10-m wind speed over China: influencing processes and sensitivities to land surface schemes. *Climate Dynamics* **50** (2018), 4189–4209.

Zhang, Q., Xu, C. Y., Gemmer, M., Chen, Y. D. & Liu, C. 2009 Changing properties of precipitation concentration in the Pearl River basin, China. *Stochastic Environmental Research* **23** (3), 377–385.

Zheng, J. Z., Wang, W. G., Cao, X. C., Feng, X. Z., Xing, W. Q., Ding, Y. M., Dong, Q. & Shao, Q. X. 2018 Responses of phosphorus use efficiency to human interference and climate change in the middle and lower reaches of the Yangtze River: historical simulation and future projections. *Journal of Cleaner Production* **201** (2018), 403–415.

Zhou, B. T. & Wang, H. J. 2006 Relationship between the boreal spring Hadley circulation and the summer precipitation in the Yangtze River valley. *Journal of Geophysical Research: Atmospheres* **111**, D16109.

First received 29 April 2021; accepted in revised form 24 November 2021. Available online 10 December 2021

doi: 10.2166/nh.2022.115

Improving BP artificial neural network model to predict the SPI in arid regions: a case study in Northern Shaanxi, China

Li Shaoxuan, Xie Jiancang, Yang Xue *, Xue Ruihua and Zhao Peiyuan
State Key Laboratory of Eco-hydraulics in Northwest Arid Region, Xi'an University of Technology, Xi'an 710048, Shaanxi, China
*Corresponding author. E-mail: xue.yang@xaut.edu.cn

YX, 0000-0003-1853-8573

ABSTRACT

Drought prediction plays an important guiding role in drought risk management. The standardized precipitation index (SPI) is a popular meteorological drought indicator to measure the degree of drought. The SPI time series is non-stationary, whereas the conventional artificial neural network (ANN) model has limitations to predict non-stationary time series. To overcome this limitation, it is essential to investigate input data preprocessing to improve the ANN model. In this paper, a hybrid model coupled with singular spectrum analysis (SSA) and backpropagation ANN is proposed (SSA-BP-ANN). The advantage of this model is that the SSA of finite-length SPI sequences does not require the adoption of boundary extensions to suppress boundary effects, while the most predictable components of the SPI can be efficiently extracted and incorporated into the model. The proposed SSA-BP-ANN model is tested in case studies at three meteorological stations in Northern Shannxi Province, China. The results show that the SSA-BP-ANN model can produce more accurate predictions than the BP-ANN model. In addition, the performance improvement of SSA on the BP-ANN model is slightly better than wavelet decomposition and empirical mode decomposition. This new hybrid prediction model has great potential for promoting drought early warning in arid regions.

Key words: BP artificial neural network, data preprocessing, decomposition hybrid model, singular spectrum analysis

HIGHLIGHTS

- Using SSA as data preprocessing tool for non-stationary time-series SPI could significantly improve the prediction performance of the BP-ANN model.
- SSA can extract more effective information from noisy time series and bring it into the prediction model.
- The SSA-BP-ANN model seems a promising method for drought early warning in arid regions.

1. INTRODUCTION

From a temporal perspective, humans have been struggling with drought for the last thousand years (Dai 2011). From a spatial perspective, drought is one of the most common natural disasters globally, characterized by high frequency, broad impact, long duration and high economic losses (Zhang *et al.* 2015b). Therefore, the research on drought has attracted much attention worldwide (Haile *et al.* 2019; Balti *et al.* 2020; Kaur & Sood 2020).

Drought generally can be divided into four categories: meteorological, agricultural, hydrological and socio-economic droughts (Kimwatu *et al.* 2021; Kolachian & Saghafian 2021; Wu *et al.* 2021; Zhang *et al.* 2021). A chronic lack of precipitation causes an imbalance in the hydrological cycle, the leading cause of the drought. Thus, many scholars have focused their attention on monitoring and preventing meteorological droughts. To quantify the severity of meteorological drought, researchers have developed many meteorological drought indicators (Waseem *et al.* 2015; Zhang *et al.* 2015a; Rehana & Naidu 2021), such as precipitation anomaly percentage (PAP) (Tahroudi *et al.* 2020), standardized precipitation index (SPI) (Mckee *et al.* 1993), standardized precipitation evapotranspiration index (SPEI) (Beguería *et al.* 2014) considering evaporation and palmer drought severity index (PDSI) (Palmer 1965). To reduce the drought influence, many different models have been carried out to predict those drought indicators (Mouatadid *et al.* 2018; Bouaziz *et al.* 2021).

SPI is one of the most commonly used meteorological drought indicators (Hayes *et al.* 2011; Bazrafshan *et al.* 2014; Qaisrani *et al.* 2021). Its calculation applies monthly rainfall data, including 1, 3, 6, 9, 12 and 24-month time scales, which facilitates the determination of drought levels and drought cycles for specific targets (Zhang *et al.* 2017). It is especially recommended when sufficient (at least 30 years) and continuous monthly rainfall data are available (Keyantash 2018). For example, Zhang *et al.* (2019a) used ARMA (autoregressive and moving average) and ARMA-GARCH (generalized autoregressive conditional heteroskedasticity) models to simulate and predict the SPI-9 drought index in Shandong Province, China, from 1965 to 2015. Moreira *et al.* (2018) validated the log-linear model and drought indices (SPI and SPEI) in modeling and predicting the drought class transitions over a sector of Eastern Europe. Durdu (2010) applied the linear model ARIMA (autoregressive integrated moving average) to predict the SPI in the Buyuk Menderes river basin, showing that the model can provide accurate results for short-term drought prediction.

The SPI time series is a typically non-linear, non-smooth series, and the existing linear time-series forecasting models are inadequate to detect the properties contained in such complex hydrological time series (Wei *et al.* 2012). Therefore, machine learning algorithms have been introduced into hydrological systems to efficiently construct predictive models, and the most widely used algorithm is artificial neural network (ANN) algorithms (Zhang *et al.* 2019b; Liu *et al.* 2020). For instance, Kousari *et al.* (2017) explored the potential of ANN in forecasting drought by using the SPI in the Fars Province of Iran. Mokhtarzad *et al.* (2017) compared the ability of ANN, adaptive neuro-fuzzy interface system (ANFIS) and support vector machine (SVM) models to predict the SPI, and the results showed that all three models have excellent predictive performance. Hassanzadeh *et al.* (2020) proved that the ANN model combined with hydrological and meteorological variables can significantly improve the ability of Iranian Tabriz to predict the SPI at different time scales. Achour *et al.*(2020) proposed a feedforward three-layer ANN model to predict the drought 2 months ahead using the SPI series at different time scales (3, 6, 9 and 12 months) and successfully helped the north-west region of Algeria cope with drought in advance.

Although ANN models have been widely applied, their capability to predict non-stationary time series is limited. The wavelet analysis (WA) shows a good effect on the preprocessing of non-stationary time series (Anshuka *et al.* 2019). Therefore, wavelet transform-based preprocessing has been largely applied with ANN models in drought prediction. For example, Belayneh *et al.* (2014) used wavelet transform to do data preprocessing on the model input, to build WA-ANN and WA-SVR (support vector regression) models for long-term SPI prediction, and the results show that the WA-ANN model has the best performance. Djerbouai & Souag-Gamane (2016) developed an ANN model based on wavelet decomposition (WD) preprocessing, and compared with the traditional random model, this new hybrid model has a higher predictive ability. Belayneh *et al.* (2016) found that wavelet transform data preprocessing tools are significantly helpful for improving the performance of machine learning prediction models. Soh *et al.* (2018) proposed two hybrid machine learning prediction models using WD as a data preprocessing (Wavelet-ARIMA-ANN and Wavelet-ANFIS) and applied them to SPEI predictions at different time scales in the Langat River Basin to obtain accurate prediction results. Khan *et al.* (2020) verified that the prediction performance of the WA-ARIMA-ANN hybrid model for the SPI is better than the single model ANN and WA-ANN.

WD preprocessing of model input data can effectively improve the predictive performance of ANN models. The decomposition of a non-stationary time series by the wavelet transform is based on the convolution of the sequence and the filter. When convolving a finite-length sequence, the information outside the sequence boundary is unknown, resulting in boundary effects on the convolution result, and additional boundary extension as required. For the sake of this, the main objective of this research is going to test the potential of singular spectrum analysis (SSA) as a data preprocessing tool for improving model prediction performance. The advantage of SSA is that it does not require any prior information and assumptions. Therefore, SSA does not require boundary extension when decomposing a finite time sequence. By combining SSA and the ANN with a backpropagation algorithm, an improved model (namely SSA-BP-ANN) is proposed to predict the SPI in this study. The proposed model will then be tested in an arid region of northern Shaanxi, China. The meteorological drought index selects the SPI of the 12-month time scale (SPI-12).

The following Section 2 introduces the general situation of the study area and the source of rainfall data. Section 3 describes the applied method, and Section 4 shows the SSA-BP-ANN model development process. Furthermore, Section 5 evaluates the performance of the model, and some main concluding remarks are presented in Section 6.

2. STUDY AREA AND DATA

The study area is the entire area under the jurisdiction of Yan'an City and Yulin City in the Loess Plateau region of northern Shaanxi Province (35°02′–39°35′N, 107°15′–110°15′E), with a total area of 79,981.9 km^2. The average altitude in the area is about 1,200 m, the annual average precipitation is 400–500 mm, the annual average temperature is 7–12 °C and the annual average evaporation is 1,200 mm. Figure 1 shows the general view of the study area.

The data used in this paper are obtained from the China Meteorological Data Network (available at http://data.cma.cn). The monthly rainfall of Hengshan, Suide and Yulin meteorological stations in Northern Shaanxi from 1955 to 2019 is selected as the data recognized and widely used in meteorological research. Table 1 summarizes the statistical characteristics of monthly rainfall at the three stations from 1955 to 2019.

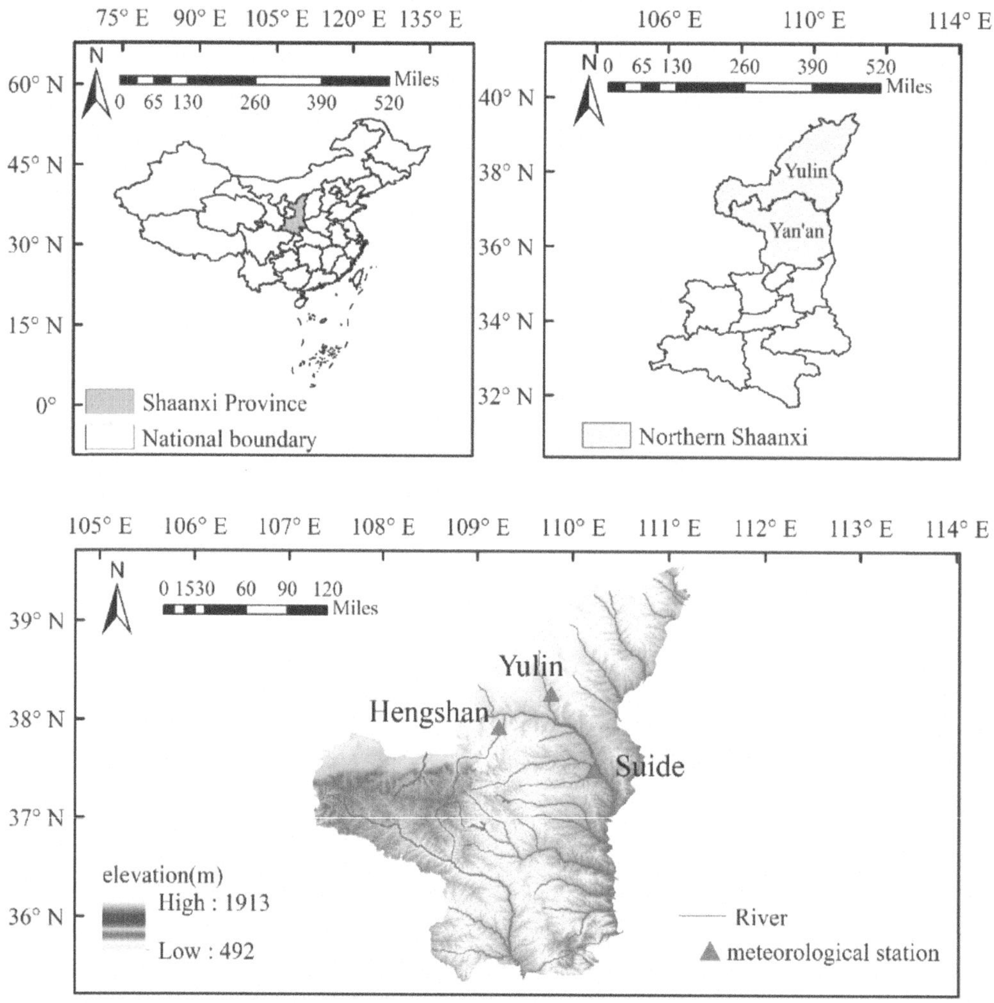

Figure 1 | Location of study areas and meteorological stations.

Table 1 | Station information and statistical characteristics of monthly rainfall data

Station	Longitude (°)	Latitude (°)	Average (mm)	Maximum (mm)	Minimum (mm)	Standard deviation (mm)
Hengshan	109.23	37.93	31.38	208.7	0	40.78
Suide	110.22	37.50	37.07	365.8	0	48.75
Yulin	109.78	38.27	33.98	329.9	0	47.53

3. METHODOLOGY

3.1. Standardized precipitation index

The SPI is widely used in meteorological drought assessment and prediction research because of its several advantages. (1) The SPI is recommended to the public by the World Meteorological Organization (Hayes *et al.* 2011); (2) the SPI only uses a long series of rainfall data, which makes drought assessment and prediction feasible in some areas where hydrological data such as evaporation are difficult to measure and (3) the SPI can evaluate and predict drought from multiple time scales (Pham *et al.* 2021).

When calculating the SPI, first select an appropriate probability distribution function to fit a long series of observed rainfall data (the most commonly used is the gamma distribution, which has been proven to be a good description of rainfall data) (Thom 1958; Chen *et al.* 2020). Secondly, calculate the cumulative probability of the observed rainfall data. Finally, the inverse normal Gaussian function with a mean value of 0 and a variance of 1 is applied to the cumulative probability to obtain the SPI (Vidyarthi & Jain 2020). The probability density function defined by the gamma distribution is shown in the following equation:

$$f(x, \alpha, \beta) = \frac{1}{\beta^{\alpha}\Gamma(\alpha)} x^{\alpha-1} e^{-x/\beta}, \text{ for } x, \alpha, \beta > 0 \tag{1}$$

where α and β represent the shape and the scale parameters, they can be obtained by maximum likelihood estimation, as shown in the following equations:

$$\hat{\alpha} = \frac{1 + \sqrt{1 + 4A/3}}{4A} \tag{2}$$

$$\hat{\beta} = \bar{x}/\bar{\alpha} \tag{3}$$

where

$$A = \ln(\bar{x}) - \frac{\sum \ln(x)}{n} \tag{4}$$

The $\Gamma(\alpha)$ is the main function in the gamma distribution, as shown in the following equation:

$$\Gamma(\alpha) = \int_0^{\infty} x^{\alpha-1} e^{-x} dx \tag{5}$$

The cumulative probability $G(x)$, then, can be obtained as follows:

$$G(x) = \int_0^{x} g(x) \tag{6}$$

Since the observed precipitation time series may contain zeros, the cumulative probability will be changed to Equations (7) and (8) (Mishra & Desai 2006):

$$H(x) = q + (1 - q)G(x) \tag{7}$$

$$q = \frac{m}{n} \tag{8}$$

where m and n are the numbers of zero precipitation and sample size, respectively. The value of SPI can be calculated according to the following equations (Lloyd-Hughes & Saunders 2002):

$$\mathrm{SPI} = -\left(k - \frac{r_0 + r_1 t + r_2 t^2}{1 + s_0 t + s_1 t^2 + s_2 t^3}\right); \; k = \sqrt{\ln \frac{1}{(H(x))^2}}; \; 0 < H(x) \leq 0.5 \tag{9}$$

$$\mathrm{SPI} = k - \frac{r_0 + r_1 t + r_2 t^2}{1 + s_0 t + s_1 t^2 + s_2 t^3}; \; k = \sqrt{\ln \frac{1}{(1 - H(x))^2}}; \; 0.5 < H(x) < 1 \tag{10}$$

where x is the monthly rainfall, $r_0 = 2.515517$, $r_1 = 0.802853$, $r_2 = 0.010328$, $s_0 = 1.432788$, $s_1 = 0.189269$ and $s_2 = 0.001308$ (Kisi *et al.* 2019). The drought threshold classification of SPI is shown in Table 2.

3.2. Backpropagation artificial neural networks

The BP-ANN is a concept proposed by scientists Rumelhart *et al.* (1986) in the 1980s. It is a multi-layer feedforward neural network trained by an error backpropagation algorithm, which generally includes an input layer, hidden layer and output layer. The neurons in each layer of this neural network only accept the input of the neurons in the previous layer, and the subsequent layers have no signal feedback to the previous layer. Each layer performs a certain transformation on the input data and then uses the output result as the input of the next layer until the final output result (Zhao *et al.* 2010).

Training the BP-ANN model is mainly divided into three steps: (1) constructing the network model, including determining the input/output data form, network level and transfer function form; (2) cyclic calculation, forward propagation calculation error and backpropagation adjustment parameters and (3) return to the trained neural network model. This paper uses a three-layer BP-ANN model to predict time series, and the network structure is shown in Figure 2.

The applied rainfall data in this study is from 1955 to 2019; thus, the SPI-12 value of 780 months can be calculated to form the entire original SPI-12 time series, which can be recorded as the set $X = \{X_1, X_2, \ldots, X_{799}, X_{780}\}$. In Figure 2, X_t (t is an integer and $1 \leq t \leq 780$) represents the SPI-12 value of the tth month of the original SPI-12 time series, X_{t-1} represents the SPI-12 value 1 month before the tth month and X_{t-n} represents the SPI-12 value n months before the tth month. The value of n will be determined by the partial autocorrelation function (PACF) in Section 4.

3.3. Singular spectrum analysis

SSA is a time-series analysis method widely used in climatology, surveying, oceanography and other fields. Its advantage lies in extracting trends, periods and noise signals from noisy time series without any prior information and gathering the most predictable components into several time series (Baydaroğlu *et al.* 2017). Therefore, it is particularly suitable for analyzing time-series data with periodic oscillations. The specific process of SSA mainly includes four steps (Yu *et al.* 2017; Mi *et al.* 2019).

Table 2 | Drought condition classifications for the SPI

Classification	SPI
Extremely wet (EW)	$\mathrm{SPI}>2.0$
Very wet (VW)	$1.50<\mathrm{SPI}\leq 2$
Moderately wet (MW)	$1.0<\mathrm{PIz}\leq 1.50$
Near normal (NN)	$-1.0<\mathrm{SPI}\leq 1.0$
Moderate drought (MD)	$-1.50<\mathrm{SPI}\leq -1.0$
Severe drought (SD)	$-2.0<\mathrm{SPI}\leq -1.50$
Extreme drought (ED)	$\mathrm{SPI}\leq -2.0$

Referring to past research on meteorological drought in northern Shaanxi, China, the time scale of SPI adopted in this paper is 12 months (SPI-12) (Wang *et al.* 2017, 2020).

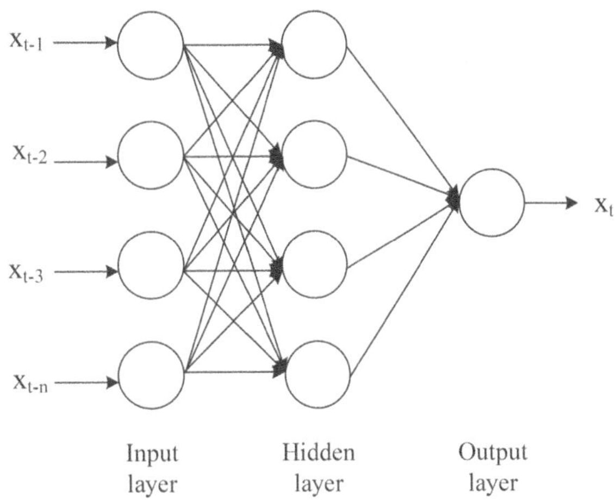

Figure 2 | Three-layer BP-ANN model structure diagram.

1. *Embedding*. For a given one-dimensional time series $X = (x_1, x_2, \cdots, x_N)$, the appropriate window length L $(2 \leqslant L \leqslant N/2)$ is chosen to construct the $L \times K$ trajectory matrix as follows:

$$X = \begin{pmatrix} x_1 & x_2 & \cdots & x_K \\ x_2 & x_3 & \cdots & x_{K+1} \\ \vdots & \vdots & & \vdots \\ x_L & x_{L+1} & \cdots & x_N \end{pmatrix} \tag{11}$$

   where $K = N - L + 1$.
2. *Singular value decomposition (SVD)*. By calculating XX^T and performing SVD, L eigenvalues can be obtained. The eigenvalues are arranged in a descending order as $\lambda_1 \geq \lambda_2 \geq \cdots \geq \lambda_L \geq 0$, and the corresponding feature vector is $\{U_1, \cdots, U_L\}$. Then, the SVD of the trajectory matrix X can be expressed as follows:

$$X = X_1 + X_2 + \cdots + X_d \tag{12}$$

$$X_i = \sqrt{\lambda_i} U_i V_i^{\mathrm{T}} (i = 1, 2, \cdots, d) \tag{13}$$

$$V_i = X^T U_i / \sqrt{\lambda_i} (i = 1, \cdots, d) \tag{14}$$

$$d = \max\{i: \lambda_i > 0\} \tag{15}$$

The elementary matrix X_i in the above formula represents the ith SVD component, and the SVD component's energy contribution to the original signal decreases as the order increases.

3. *Grouping*. Dividing the subscript interval $\{1, 2, \cdots, d\}$ of the elementary matrix X_i into P unconnected subsets $\{I_1, I_2, \cdots, I_p\}$, denoted as $I = \{i_1, i_2, \cdots, i_m\}$, so the composite matrix $X_I = X_{i1} + X_{i2} + \cdots + X_{im}$. Calculating each composite matrix of $I = I_1, I_2, \cdots, I_p$, then Equation (16) can be obtained as follows:

$$X = X_{I1} + X_{12} + \cdots + X_{Ip} \tag{16}$$

   Among them, the selection process of $\{I_1, I_2, \cdots, I_p\}$ is the grouping.
4. *Refactor*. By converting the matrix obtained by grouping in step (3) into a series of new reconstruction components of length N. Define $Y = X_{Ik}$, $\{y_1, y_2, \cdots, y_N\}$ is the sequence obtained by Y through diagonal averaging. Let

$L^* = min\{L, K\}$, $K^* = max\{L, K\}$ and $N = K + L - 1$. The diagonal average formula is expressed as follows:

$$y_k = \begin{cases} \frac{1}{k}\sum_{p=1}^{k} y^*_{p,k-p+1} & 1 \leqslant k \leqslant L^* \\ \frac{1}{L^*}\sum_{p=1}^{L^*} y^*_{p,k-p+1} & L^* \leq k \leq K^* \\ \frac{1}{N-k+1}\sum_{p=k-K^*+1}^{N-K^*+1} y^*_{p,k-p+1} & K^* \leqslant k \leqslant N \end{cases} \quad (17)$$

Through diagonal averaging, the grouped matrix can be converted into its corresponding time-series data. The time-series data of each group show the hidden characteristics of the original sequence, such as long-term trend, periodicity and noise signal. The value of the original time series is equal to the sum of the characteristic decomposition series, as shown in the following equation:

$$\boldsymbol{x}_n = \sum_{k=1}^{p} f_n^{(k)} (n = 1, 2, \cdots, N) \quad (18)$$

where the sequence $f_n^{(k)}$ corresponding to each k in the formula is the diagonal mean value of the matrix X_{Ik}.

3.4. SSA-BP-ANN prediction model

The SSA-BP-ANN prediction model is a new hybrid prediction model, and its specific prediction process is shown in Figure 3. The whole process can be roughly divided into four steps: (1) the division of training set and testing set for the calculated SPI-12 time series, where the training set accounts for 70%; (2) the SSA method is used to decompose the training set, the

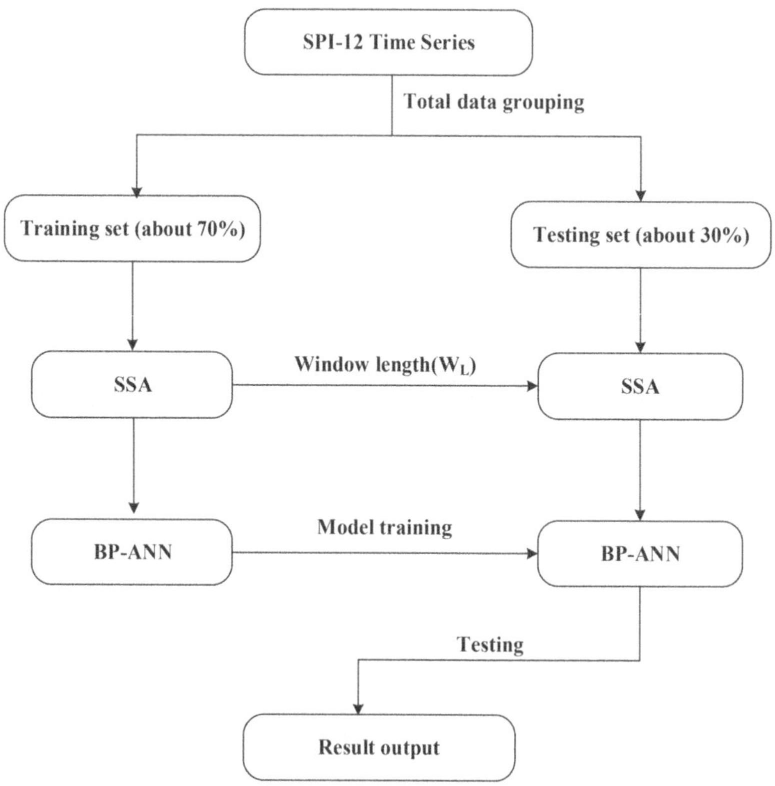

Figure 3 | Flowchart of SSA-BP-ANN model prediction.

parameter (window length, W_L) of the method is determined in the decomposition process and then the testing set is decomposed according to the same W_L; (3) the decomposed training set is fed into the BP-ANN model for model training and (4) the decomposed testing set is fed into the trained BP-ANN model to predict SPI-12 time series, and the evaluation indices are used to evaluate the performance of the model.

3.5. Model performance evaluation

In the stage of evaluating the performance of the model, one cannot rely on the results of a single index. A series of different evaluation indexes should be selected for comprehensive evaluation from multiple angles (Dawson *et al.* 2007). In this study, the evaluation indices include the mean absolute error (MAE), root-mean-square error (RMSE), coefficient of determination (R^2) and Nash–Sutcliffe efficiency coefficient (NSE), as shown in the following equations:

$$\mathrm{MAE} = \sum_{i=1}^{N} \frac{|\hat{y}_t - y_i|}{N} \tag{19}$$

$$\mathrm{RMSE} = \sqrt{\frac{\sum_{i=1}^{N} (\hat{y}_t - y_i)^2}{N}} \tag{20}$$

$$R^2 = \frac{\sum_{i=1}^{N} (\hat{y}_t - \overline{y_t})}{\sum_{i=1}^{N} (y_i - \overline{y_t})^2} \tag{21}$$

$$\mathrm{NSE} = 1 - \frac{\sum_{i=1}^{n} (y_i - \hat{y}_t)^2}{\sum_{i=1}^{n} (y_i - \overline{y_t})^2} \tag{22}$$

$$\overline{y_t} = \frac{1}{N} \sum_{i=1}^{N} y_i \tag{23}$$

where N is the number of samples, $\hat{y}_t$ is the predicted value, y_i is the observed value and $\overline{y_t}$ is the mean of the observed value (Fung *et al.* 2019).

In general, the performance indicators MAE (MAE$\geq$0) and RMSE (RMSE$\geq$0) are used to reflect the actual situation of the model prediction error. The smaller their values, the better the accuracy of the prediction model in describing the experimental data. The R^2 ($0 \leq R^2 \leq 1$) and NSE (NSE$\leq$1) are used to measure the fitting accuracy of the prediction model. The larger their values, the better the performance of the model and the higher the fitting accuracy.

These indices are not only used to evaluate the performance of the SSA-BP-ANN model, but also used to compare the differences in model performance improvement of different data processing methods.

4. SSA-BP-ANN MODEL DEVELOPMENT

4.1. SPI-12 calculation

As the prediction object, the first step is to calculate the time series of SPI-12 based on the observed rainfall data. Figure 4 shows the results of SPI-12 time series for different meteorological stations. This processing follows: (1) applying the χ^2 and Kolgomorov–Smirnov (K–S) tests to detect whether the monthly rainfall series follows the gamma distribution; (2) using Matlab code to calculate the original SPI-12 time series (code available at https://www.csdn.net/) and (3) dividing the total data into two sets: about 70% as the training set (550) and about 30% as the testing set (230).

4.2. Singular spectrum decomposition of the training set and testing set

The object of singular spectral decomposition (Bonizzi *et al.* 2014) is the training set and testing set data. The purpose of this is to reduce the influence of noise, extract more predictable information and do data preprocessing for the input of subsequent models. Two parameters need to be set when using the SSA method, namely W_L and the number of groups (m). If the value of

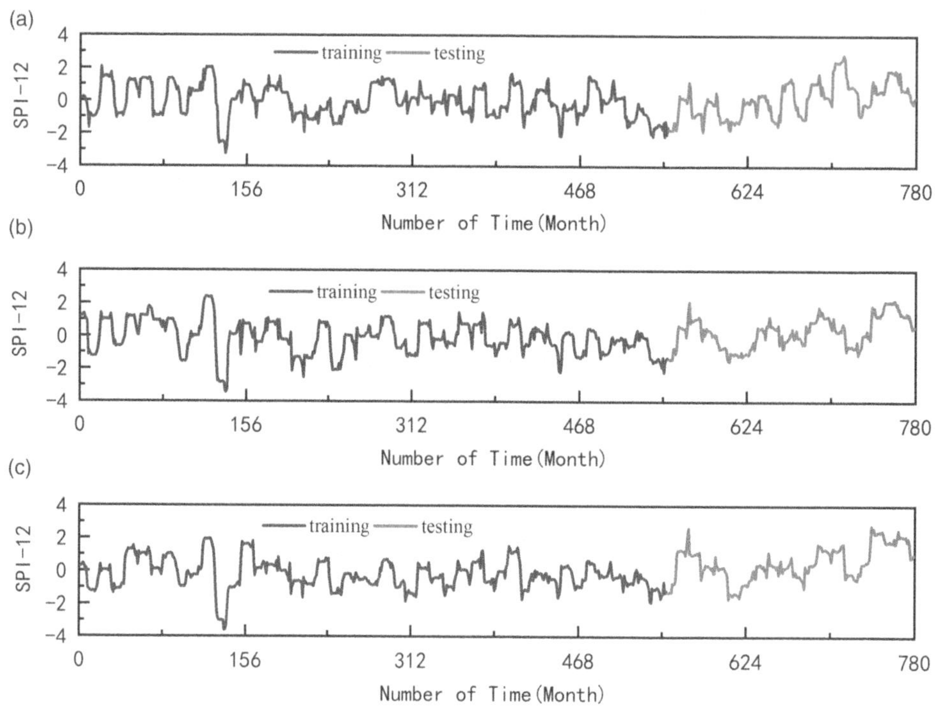

Figure 4 | SPI-12 time series in (a) Suide, (b) Hengshan and (c) Yulin stations.

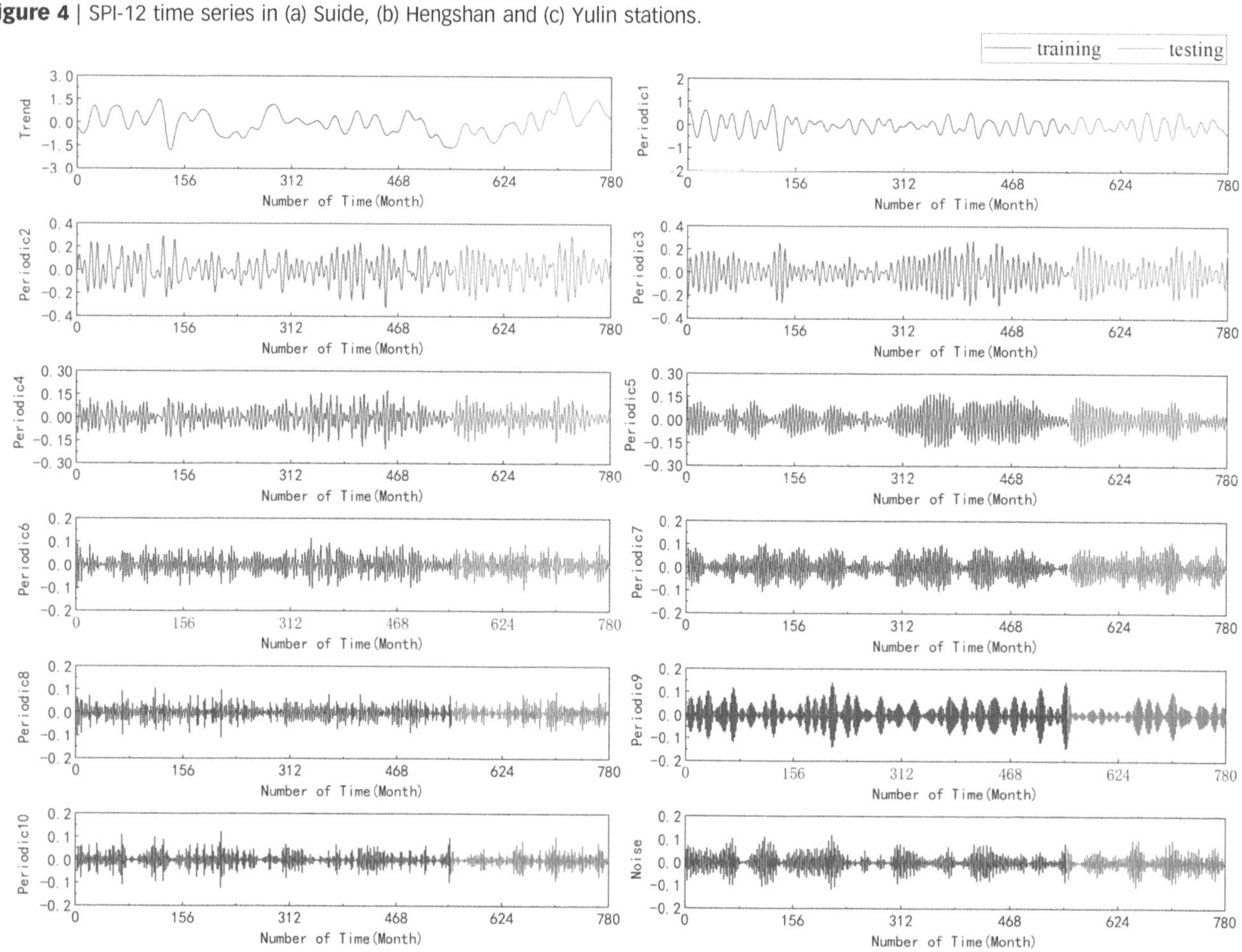

Figure 5 | Singular spectrum decomposition of SPI-12 time series at the Suide station.

the W_L is small (e.g., $W_L \leq 20$), m is meaningless, because the grouping may hide some useful information (Zuo *et al.* 2020). In this work, W_L is set to be 12 because the period of SPI-12 is 12 months. Therefore, the original SPI-12 time series will be decomposed and ordered into 12 subsequences. After determining the appropriate W_L, the training set is first decomposed to verify the reliability of W_L. If W_L is reasonable, the same W_L is used to decompose the testing set. This approach can effectively avoid bringing the future information of the testing set into the training process of the model.

For the subsequences after singular spectral decomposition, we would like to name the first subsequence as the trend component, the last subsequence as the noise component and all the intermediate subsequences as the periodic component. From Figures 5–7, the original SPI-12 time series of three stations are all decomposed into 1 trend component, 10 periodic components and 1 noise component. The trend component has the largest amplitude, the lowest frequency and the longest wavelength, which roughly reflects the changing trend of the original SPI-12 time series. The amplitude of the periodic components 1–10 gradually decreases, the frequency gradually increases and the wavelength gradually decreases, reflecting the fluctuation of the original SPI-12 time series. Each subsequence shows the characteristics of the original sequence.

4.3. BP-ANN model development

The BP-ANN model is developed in two steps. First, the forecasting strategy is determined for forecasting 1 month ahead SPI using the past SPI data, namely 'decomposition-forecast-reconstruction', and the process is shown in Figure 8.

From Figure 8, this strategy predicts each component separately and finally adds the prediction results of each component to obtain the predicted value of the original SPI-12 sequence. The input variables for the BP-ANN model were identified based on PACF values between the component of the training set at the time step t and the corresponding lagged components. The PACF of SPI-12 components in the training set of three stations is shown in Figures 9–11.

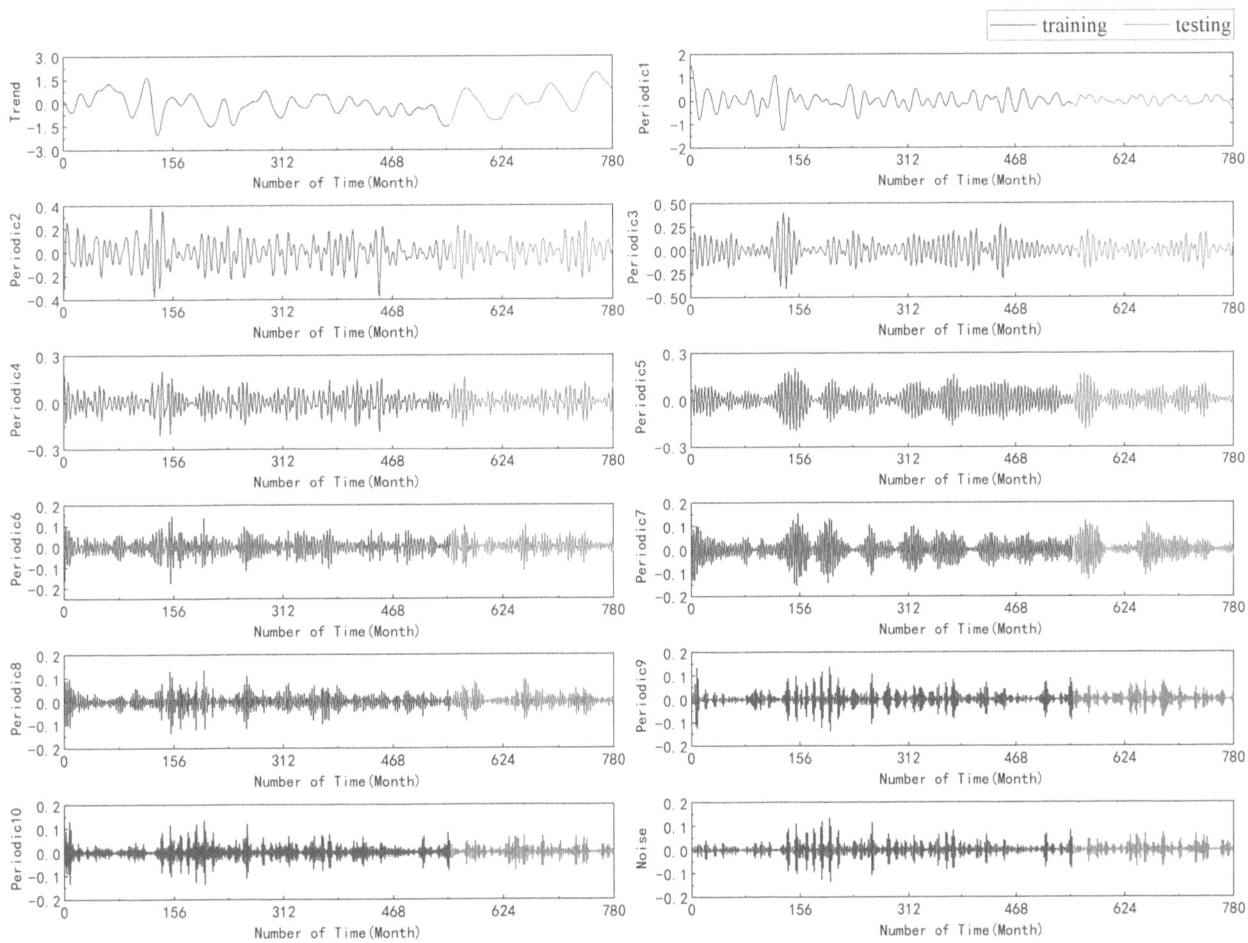

Figure 6 | Singular spectrum decomposition of SPI-12 time series at the Hengshan station.

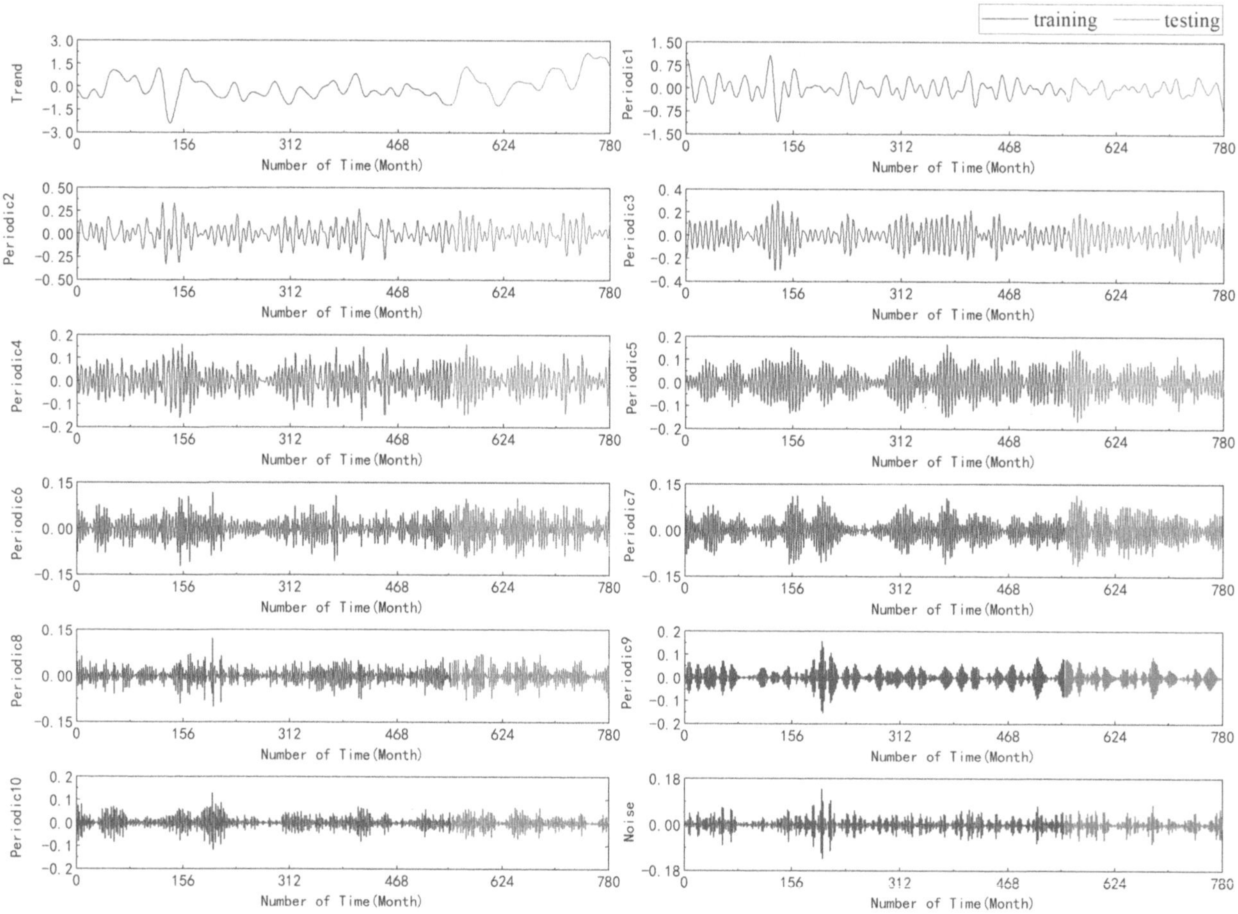

Figure 7 | Singular spectrum decomposition of SPI-12 time series at the Yulin station.

By drawing the PACF diagram of each component of SPI-12, find out the lag factor outside the confidence interval (blue line), which is the predictor. For example, for the trend component in Figure 9, the number of lags outside the confidence interval is 1, 2 and 3. Therefore, these three lags are selected as the input variables of the model, and the same number of lags is selected as the input variables of this component in the testing set. Based on the 'decomposition-forecast-reconstruction' strategy, the predictors of all components selected by the PACF are shown in Table 3. In Table 3, the SPI-12 time series of three stations are all decomposed into 12 components, and each component has the same length as the original time series. Therefore, each component can also be expressed in the form of a set $X = \{X_1, X_2, \ldots, X_{799}, X_{780}\}$. As described in Section 3, X_t (t is an integer and $1 \le t \le 780$) represents the tth value in the set X. Using the trend component of the Suide station in Table 3 as an example, the model structure shown in Figure 2 and the value of n determined in Figure 9 ($n = 3$) are used. Then the input layer of the BP-ANN model has three input variables, X_{t-1}, X_{t-2} and X_{t-3}, and the output layer is X_t.

In the second step, using the training set data to train the BP-ANN model to get the best network structure, and bring in the testing set data to get the final prediction result. The BP-ANN model developed in this research is composed of three layers. At present, the input layer and output layer of each SPI-12 component have been determined. Next, the optimal number of neurons in the hidden layer is determined by using the trial and error method. To this end, the number of neurons in the hidden layer ranges from 1 to 15, and the most commonly used Levenberg–Marquardt (LM) algorithm is used to train the neural network (Hagan & Menhaj 1994).

In the development process of the model, the transfer function of the hidden layer is a tangent sigmoid function (tansig), and the transfer function of the output layer is a linear function (purelin). The model with the minimum performance standards MAE and RMSE is selected as the best network structure. The model parameters are set as follows: training times are 10,000, the learning rate is 0.01, the minimum error of training target is 0.01 and the maximum number of confirmation failures is 6. Therefore, the optimal network structure of all components of the three stations is shown in Table 4.

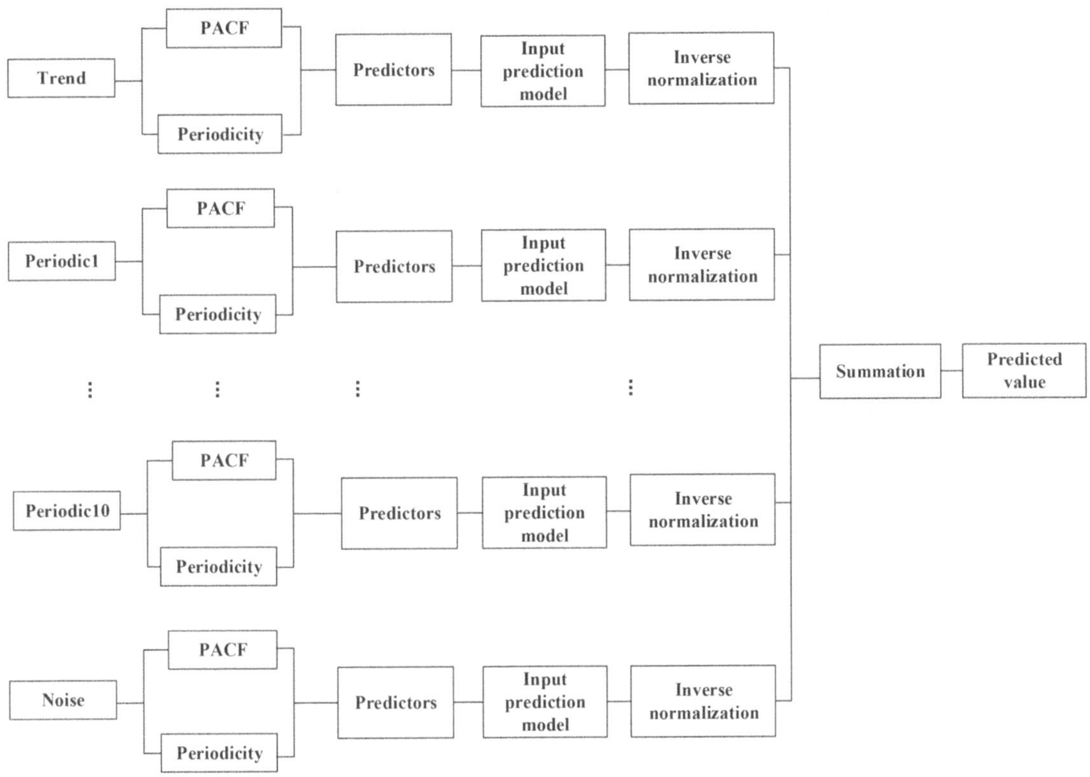

Figure 8 | Flowchart of decomposition–forecast–reconfiguration.

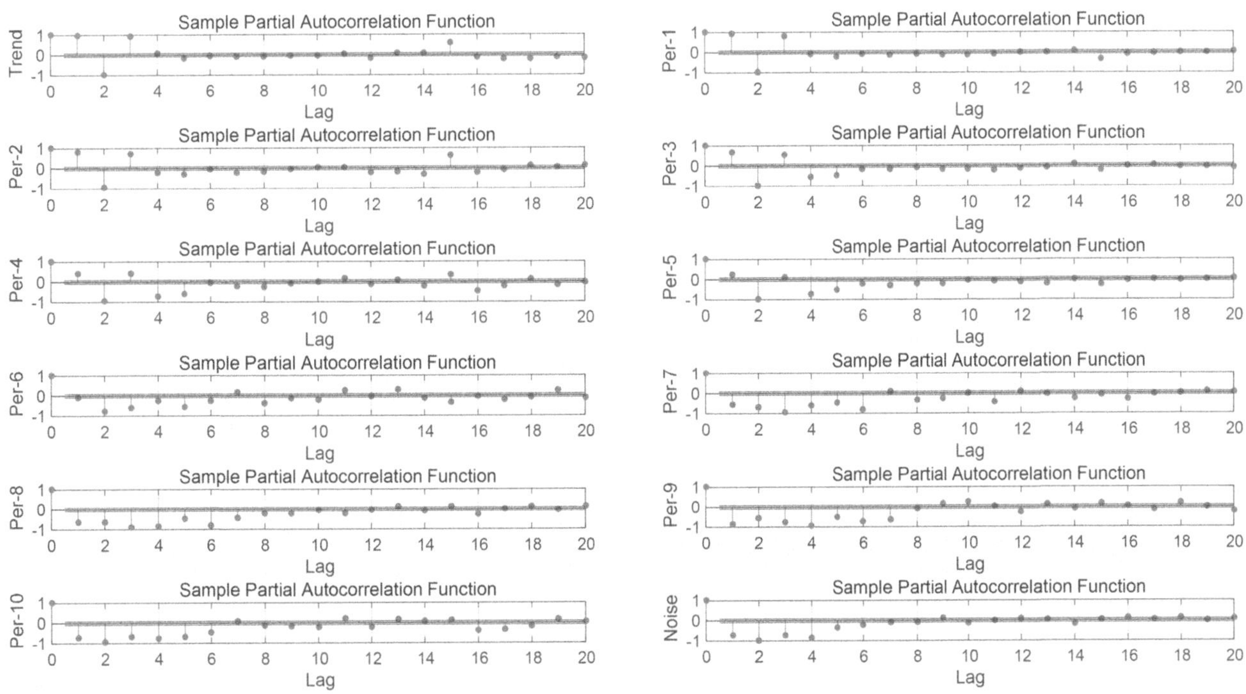

Figure 9 | PACF diagram of SPI-12 component in training set at the Suide station. Please refer to the online version of this paper to see this figure in color: http://dx.doi.org/10.2166/nh.2022.115.

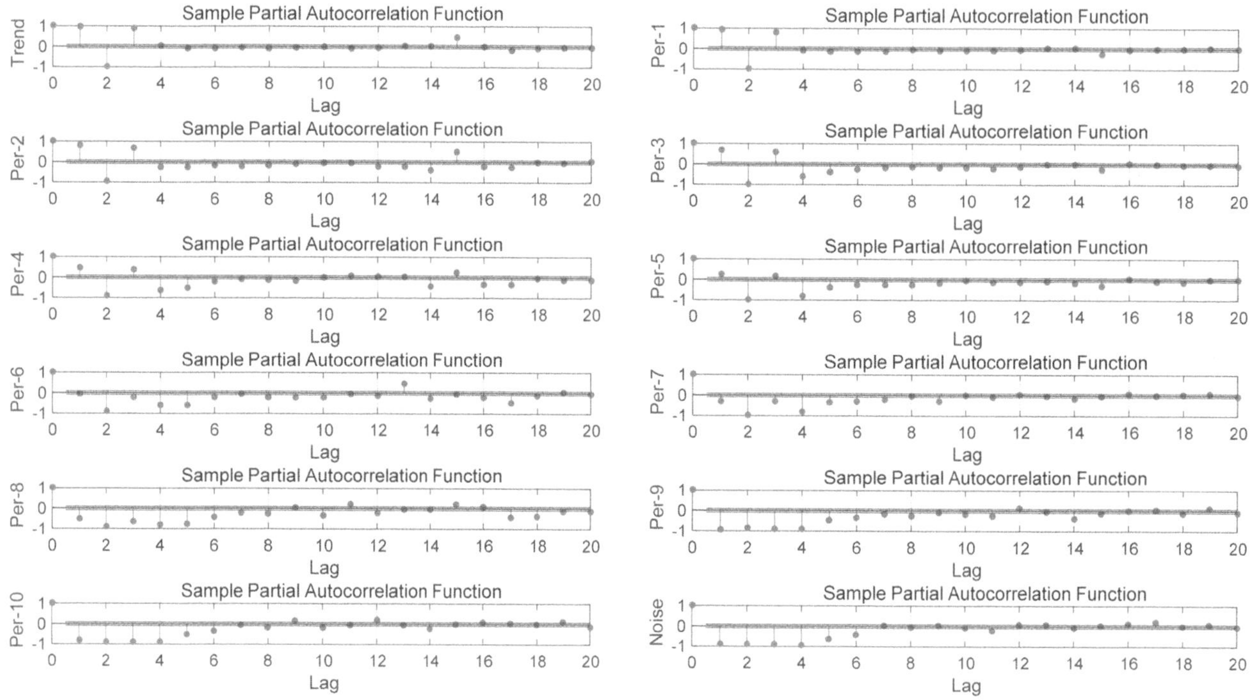

Figure 10 | PACF diagram of SPI-12 component in training set at the Hengshan station.

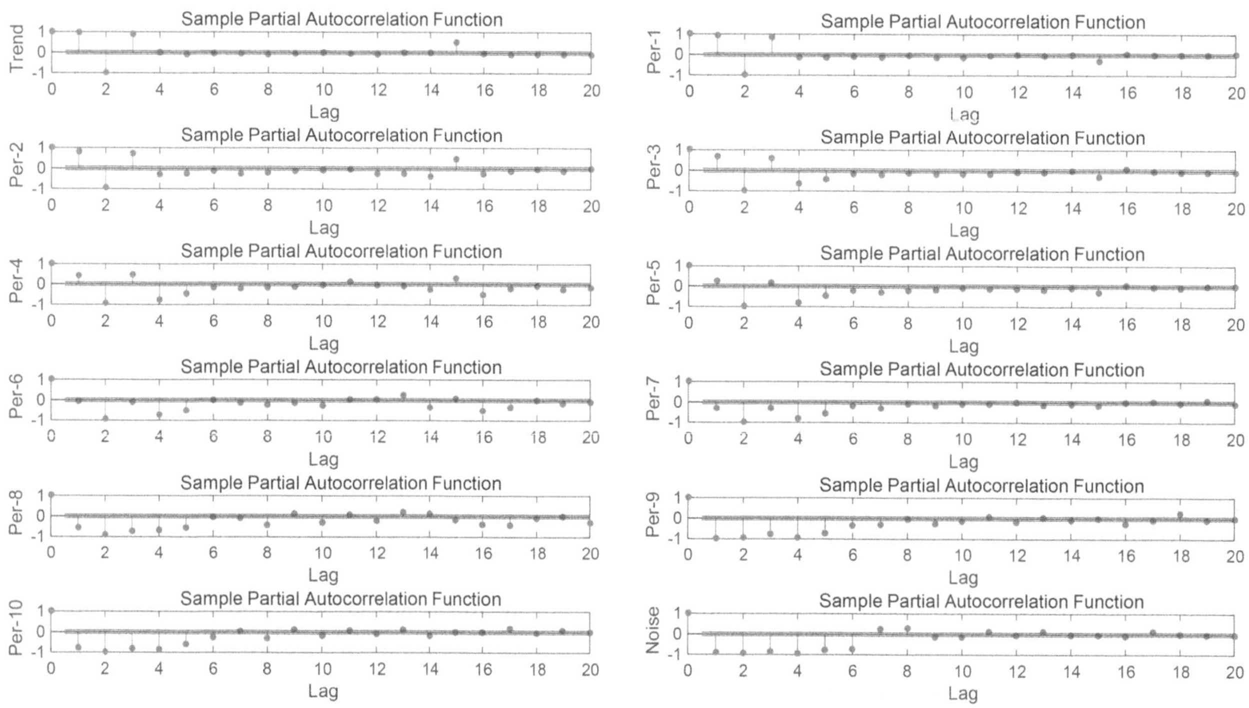

Figure 11 | PACF diagram of SPI-12 component in training set at the Yulin station.

5. RESULTS AND DISCUSSIONS

5.1. Assessment of SSA-BP-ANN model performance

The performance of both BP-ANN and SSA-BP-ANN models is assessed statistically by the evaluation indices as shown above. The corresponding results are present in Table 5 as values and in Figure 12 as a Taylor diagram.

Table 3 | Predictors selected by the PACF

Stations	SSA component	Model input variables	Model output
Suide	Trend	$\{X_{t-1}, X_{t-2}, X_{t-3}\}$	X_t
	Periodic-1	$\{X_{t-1}, X_{t-2}, X_{t-3}\}$	X_t
	Periodic-2	$\{X_{t-1}, X_{t-2}, X_{t-3}\}$	X_t
	Periodic-3	$\{X_{t-1}, X_{t-2}, X_{t-3}, X_{t-4}, X_{t-5}\}$	X_t
	Periodic-4	$\{X_{t-1}, X_{t-2}, X_{t-3}, X_{t-4}, X_{t-5}\}$	X_t
	Periodic-5	$\{X_{t-1}, X_{t-2}, X_{t-3}, X_{t-4}, X_{t-5}\}$	X_t
	Periodic-6	$\{X_{t-1}, X_{t-2}, X_{t-3}, X_{t-4}, X_{t-5}\}$	X_t
	Periodic-7	$\{X_{t-1}, X_{t-2}, X_{t-3}, X_{t-4}, X_{t-5}\}$	X_t
	Periodic-8	$\{X_{t-1}, X_{t-2}, X_{t-3}, X_{t-4}, X_{t-5}, X_{t-6}\}$	X_t
	Periodic-9	$\{X_{t-1}, X_{t-2}, X_{t-3}, X_{t-4}, X_{t-5}, X_{t-6}\}$	X_t
	Periodic-10	$\{X_{t-1}, X_{t-2}, X_{t-3}, X_{t-4}, X_{t-5}, X_{t-6}\}$	X_t
	Noise	$\{X_{t-1}, X_{t-2}, X_{t-3}, X_{t-4}, X_{t-5}\}$	X_t
Hengshan	Trend	$\{X_{t-1}, X_{t-2}, X_{t-3}\}$	X_t
	Periodic-1	$\{X_{t-1}, X_{t-2}, X_{t-3}\}$	X_t
	Periodic-2	$\{X_{t-1}, X_{t-2}, X_{t-3}\}$	X_t
	Periodic-3	$\{X_{t-1}, X_{t-2}, X_{t-3}, X_{t-4}, X_{t-5}\}$	X_t
	Periodic-4	$\{X_{t-1}, X_{t-2}, X_{t-3}, X_{t-4}, X_{t-5}\}$	X_t
	Periodic-5	$\{X_{t-1}, X_{t-2}, X_{t-3}, X_{t-4}, X_{t-5}\}$	X_t
	Periodic-6	$\{X_{t-1}, X_{t-2}, X_{t-3}, X_{t-4}, X_{t-5}\}$	X_t
	Periodic-7	$\{X_{t-1}, X_{t-2}, X_{t-3}, X_{t-4}, X_{t-5}\}$	X_t
	Periodic-8	$\{X_{t-1}, X_{t-2}, X_{t-3}, X_{t-4}, X_{t-5}, X_{t-6}\}$	X_t
	Periodic-9	$\{X_{t-1}, X_{t-2}, X_{t-3}, X_{t-4}, X_{t-5}, X_{t-6}\}$	X_t
	Periodic-10	$\{X_{t-1}, X_{t-2}, X_{t-3}, X_{t-4}, X_{t-5}, X_{t-6}\}$	X_t
	Noise	$\{X_{t-1}, X_{t-2}, X_{t-3}, X_{t-4}, X_{t-5}, X_{t-6}\}$	X_t
Yulin	Trend	$\{X_{t-1}, X_{t-2}, X_{t-3}\}$	X_t
	Periodic-1	$\{X_{t-1}, X_{t-2}, X_{t-3}\}$	X_t
	Periodic-2	$\{X_{t-1}, X_{t-2}, X_{t-3}\}$	X_t
	Periodic-3	$\{X_{t-1}, X_{t-2}, X_{t-3}, X_{t-4}, X_{t-5}\}$	X_t
	Periodic-4	$\{X_{t-1}, X_{t-2}, X_{t-3}, X_{t-4}, X_{t-5}\}$	X_t
	Periodic-5	$\{X_{t-1}, X_{t-2}, X_{t-3}, X_{t-4}, X_{t-5}\}$	X_t
	Periodic-6	$\{X_{t-1}, X_{t-2}, X_{t-3}, X_{t-4}, X_{t-5}\}$	X_t
	Periodic-7	$\{X_{t-1}, X_{t-2}, X_{t-3}, X_{t-4}, X_{t-5}\}$	X_t
	Periodic-8	$\{X_{t-1}, X_{t-2}, X_{t-3}, X_{t-4}, X_{t-5}\}$	X_t
	Periodic-9	$\{X_{t-1}, X_{t-2}, X_{t-3}, X_{t-4}, X_{t-5}\}$	X_t
	Periodic-10	$\{X_{t-1}, X_{t-2}, X_{t-3}, X_{t-4}, X_{t-5}, X_{t-6}\}$	X_t
	Noise	$\{X_{t-1}, X_{t-2}, X_{t-3}, X_{t-4}, X_{t-5}, X_{t-6}\}$	X_t

From Table 5, the fitting accuracy of the SSA-BP-ANN model is significantly improved than that of the BP-ANN model, and the prediction errors are reduced to varying degrees. At the Suide station, the SSA-BP-ANN model produces R^2 as 0.86, NSE as 0.86, both 18% higher compared to the BP-ANN model, and the MAE and the RMSE are 33 and 39% lower. At the Hengshan station, the SSA-BP-ANN model produces R^2 as 0.88, NSE as 0.87, both 13 and 12% higher compared to the BP-ANN model, and the MAE and the RMSE are 28 and 23% lower. At the Yulin station, the SSA-BP-ANN model produces R^2 as 0.88, NSE as 0.86, both 14 and 18% higher compared to the BP-ANN model, and the MAE and the RMSE are 34 and 27% lower.

Figure 12 displays the correlation coefficient, root mean square difference (RMSD) and standard deviation between the predicted and observed values in the testing set of three stations. According to Figure 12, the general performance of SSA-BP-ANN provides more accurate prediction results than the BP-ANN model. For example, the correlation coefficients of the SSA-BP-ANN model are all higher than the BP-ANN model, which are closer to 0.95. On the other hand, the RMSD values from the SSA-BP-ANN model are all lower than the BP-ANN model, which are closer to 0. These results above indicate the proposed approach to data preprocessing could significantly improve the performance of the model, which agrees with the conclusions from other studies (Khan *et al.* 2018; Mehdizadeh *et al.* 2020; Hinge & Sharma 2021).

Figure 13 shows the observed and predicted time series for the testing set at the three stations. In general, the fitting from the BP-ANN model is significantly worse than the SSA-BP-ANN model, especially when predicting the valley value of the

Table 4 | Network architecture of the best BP-ANN model for all components

Stations	SSA component	Network architecture	Training algorithm	Hidden transfer function	Output transfer function
Suide	Trend	3-9-1	Trainlm	Tansig	Purelin
	Periodic-1	3-9-1	Trainlm	Tansig	Purelin
	Periodic-2	3-3-1	Trainlm	Tansig	Purelin
	Periodic-3	5-4-1	Trainlm	Tansig	Purelin
	Periodic-4	5-6-1	Trainlm	Tansig	Purelin
	Periodic-5	5-4-1	Trainlm	Tansig	Purelin
	Periodic-6	5-3-1	Trainlm	Tansig	Purelin
	Periodic-7	5-10-1	Trainlm	Tansig	Purelin
	Periodic-8	6-3-1	Trainlm	Tansig	Purelin
	Periodic-9	6-1-1	Trainlm	Tansig	Purelin
	Periodic-10	6-10-1	Trainlm	Tansig	Purelin
	Noise	5-5-1	Trainlm	Tansig	Purelin
Hengshan	Trend	3-8-1	Trainlm	Tansig	Purelin
	Periodic-1	3-8-1	Trainlm	Tansig	Purelin
	Periodic-2	3-10-1	Trainlm	Tansig	Purelin
	Periodic-3	5-6-1	Trainlm	Tansig	Purelin
	Periodic-4	5-9-1	Trainlm	Tansig	Purelin
	Periodic-5	5-6-1	Trainlm	Tansig	Purelin
	Periodic-6	5-1-1	Trainlm	Tansig	Purelin
	Periodic-7	5-1-1	Trainlm	Tansig	Purelin
	Periodic-8	6-5-1	Trainlm	Tansig	Purelin
	Periodic-9	6-6-1	Trainlm	Tansig	Purelin
	Periodic-10	6-7-1	Trainlm	Tansig	Purelin
	Noise	6-10-1	Trainlm	Tansig	Purelin
Yulin	Trend	3-3-1	Trainlm	Tansig	Purelin
	Periodic-1	3-1-1	Trainlm	Tansig	Purelin
	Periodic-2	3-3-1	Trainlm	Tansig	Purelin
	Periodic-3	5-9-1	Trainlm	Tansig	Purelin
	Periodic-4	5-2-1	Trainlm	Tansig	Purelin
	Periodic-5	5-3-1	Trainlm	Tansig	Purelin
	Periodic-6	5-9-1	Trainlm	Tansig	Purelin
	Periodic-7	5-9-1	Trainlm	Tansig	Purelin
	Periodic-8	5-7-1	Trainlm	Tansig	Purelin
	Periodic-9	5-9-1	Trainlm	Tansig	Purelin
	Periodic-10	6-5-1	Trainlm	Tansig	Purelin
	Noise	7-2-1	Trainlm	Tansig	Purelin

Table 5 | Performance statistics of prediction models for three stations

Stations	Models	R^2	NSE	MAE	RMSE
Suide	BP-ANN	0.73	0.73	0.36	0.55
	SSA-BP-ANN	0.86	0.86	0.24	0.39
Hengshan	BP-ANN	0.78	0.78	0.29	0.43
	SSA-BP-ANN	0.88	0.87	0.21	0.33
Yulin	BP-ANN	0.77	0.73	0.32	0.48
	SSA-BP-ANN	0.88	0.86	0.21	0.35

time series. From Figure 13, nearly one-third of the predicted results time series is lower than the valley values of the observed values. On the other hand, the SSA-BP-ANN model is more accurate in predicting peaks and valleys, since the time-shift error is significantly lower than the BP-ANN model. This difference confirms the importance of the data preprocessing method for BP-ANN model training.

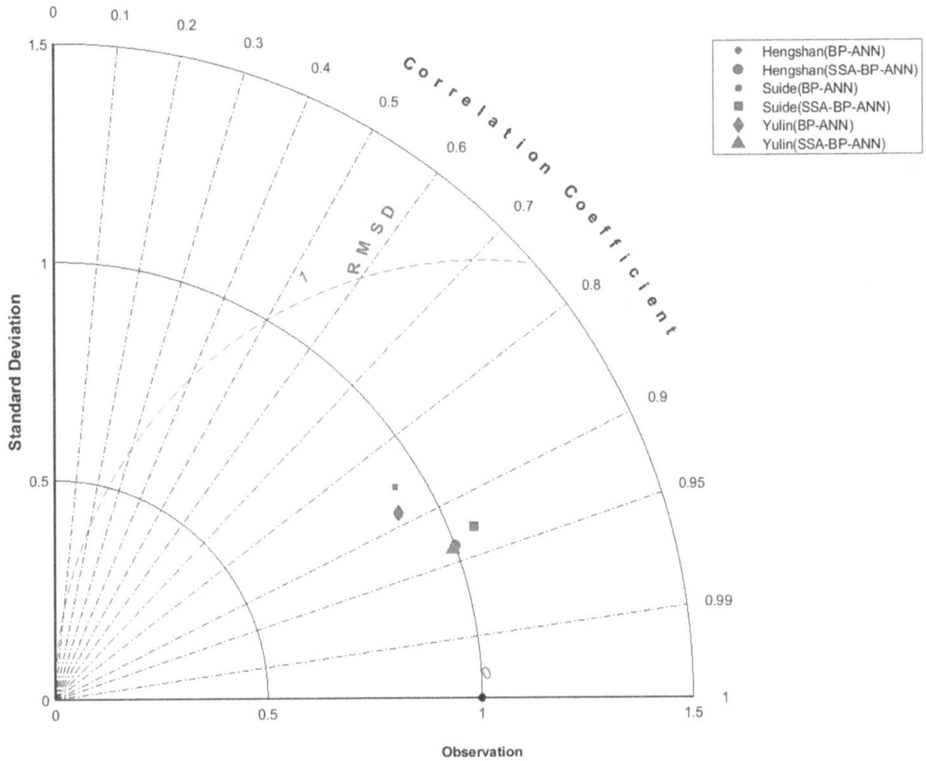

Figure 12 | Taylor diagram of model's errors for all stations in the testing set.

5.2. Comparison between data processing methods

The SSA data processing method used in this paper is a new attempt in the prediction of drought by the ANN model. However, currently the most widely used data processing methods are WD and empirical mode decomposition (EMD), and we have also conducted experiments to compare these three methods. Their detailed principles can be obtained from Özger *et al.* (2020). Here, we directly present their performance by the evaluation indices applied in Table 6.

From Tables 5 and 6, the fitting accuracy of the BP-ANN model displays significant improvement after data processing, and the R^2 and NSE of all hybrid models are above 0.85. Meanwhile, the three data processing methods show almost identical prediction performance in terms of various statistical indicators. This indicates that the SSA method has great potential in improving model performance.

Meanwhile, the Kruskal–Wallis (KW) test has been done on each forecast time series to evaluate the statistical significance of all models (Başakın *et al.* 2021), and the result is shown in Table 7.

From Table 7, the H_0 hypothesis was rejected for all three hybrid models in the KW test, indicating that there is no significant difference between the means of predicted and observed values of the model and that the model accuracy was high. On the other hand, there is a significant difference between the means of predicted and observed values of the BP-ANN model at the Hengshan and Yulin stations, and the model accuracy is low. Therefore, the three SSA, EMD and WD methods have improved the reliability and accuracy of model prediction results to a certain extent.

To compare these methods more intuitively, we also produced the Taylor diagram (Figure 14) and a box plot (Figure 15) to present the evaluation results.

From Figure 14, three hybrid models provide high prediction accuracy, and the differences between the different data processing methods are neglectable. The correlation coefficients of the three hybrid models are all higher than the BP-ANN model and close to 0.95.

From Figure 15, the three hybrid models are overall closer to 0 in terms of the absolute value of prediction error (AVE) than the BP-ANN model, and the variation range is slightly lower than the BP-ANN model. It again confirms the notion that the data processing method can improve the prediction performance of the model. Among them, the average value of the AVE of

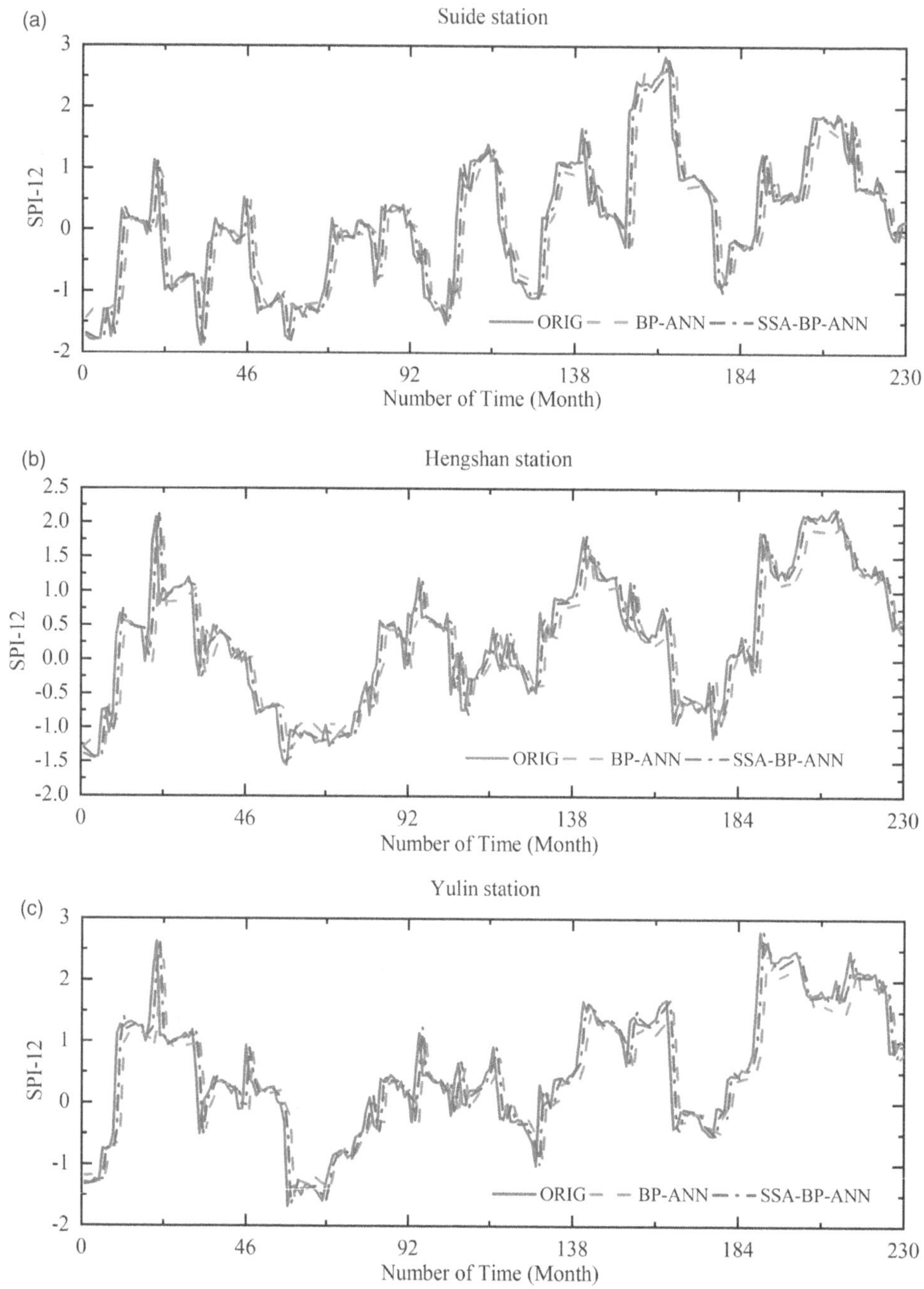

Figure 13 | Time-series plot of predicted results for all stations in the testing set.

Table 6 | Performance statistics of prediction models for three stations

Stations	Models	R^2	NSE	MAE	RMSE
Suide	EMD-BP-ANN	0.87	0.88	0.13	0.37
	WD-BP-ANN	0.86	0.86	0.26	0.40
Hengshan	EMD-BP-ANN	0.87	0.88	0.23	0.33
	WD -BP-ANN	0.88	0.88	0.21	0.33
Yulin	EMD-BP-ANN	0.89	0.89	0.23	0.34
	WD -BP-ANN	0.88	0.88	0.21	0.35

Table 7 | Kruskal–Wallis test at 95% significance level

Stations	Models	P-value	*H_0
Suide	BP-ANN	0.7252	*Reject*
	SSA-BP-ANN	0.9134	*Reject*
	EMD-BP-ANN	0.9034	*Reject*
	WD-BP-ANN	0.9262	*Reject*
Hengshan	BP-ANN	0.4597	Accept
	SSA-BP-ANN	0.8978	*Reject*
	EMD-BP-ANN	0.7247	*Reject*
	WD-BP-ANN	0.9017	*Reject*
Yulin	BP-ANN	0.3021	Accept
	SSA-BP-ANN	0.8202	*Reject*
	EMD-BP-ANN	0.8862	*Reject*
	WD-BP-ANN	0.8109	*Reject*

*H_0 represents a hypothesis based on the significant difference between mean predicted and observed values.

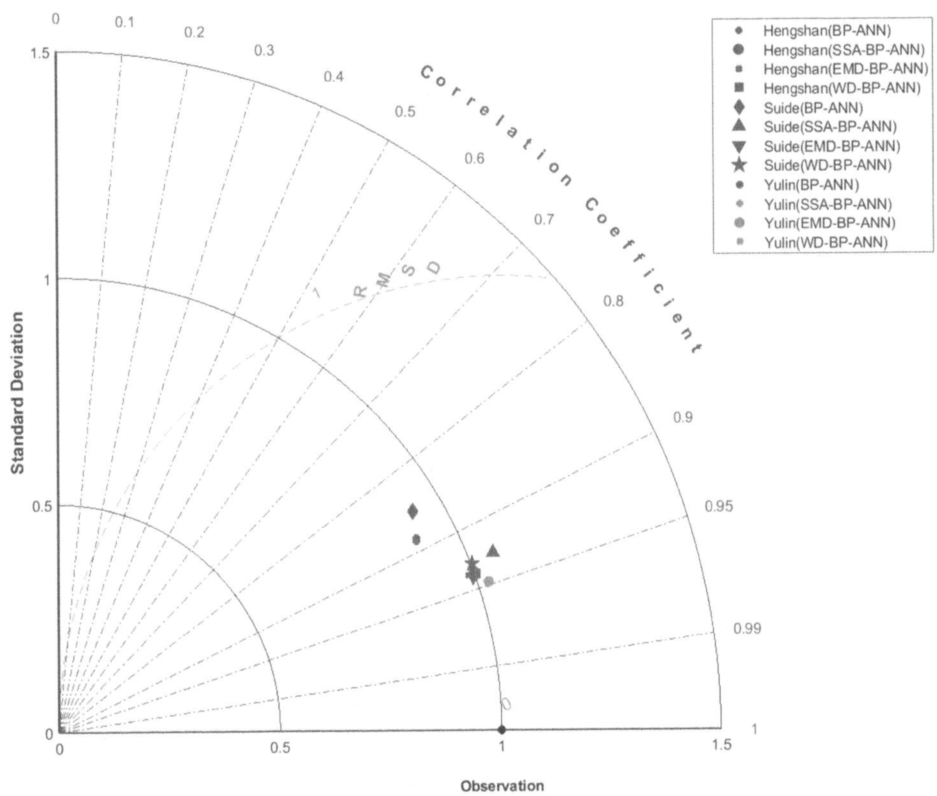

Figure 14 | Taylor diagram of four models' errors for all stations in the testing set.

the SSA method is the smallest over three stations, indicating that the SSA method is slightly better than the EMD and WD methods in SPI prediction.

6. CONCLUSION

In this study, a prediction model coupled with SSA and BP-ANN was established and applied to the prediction of SPI-12 1 month ahead at three stations in northern Shaanxi, China. The following conclusions can be drawn from the study:

1. The result validates that by preprocessing the singular spectrum decomposition of the input data of the BP-ANN model, the prediction performance of the model is greatly improved.

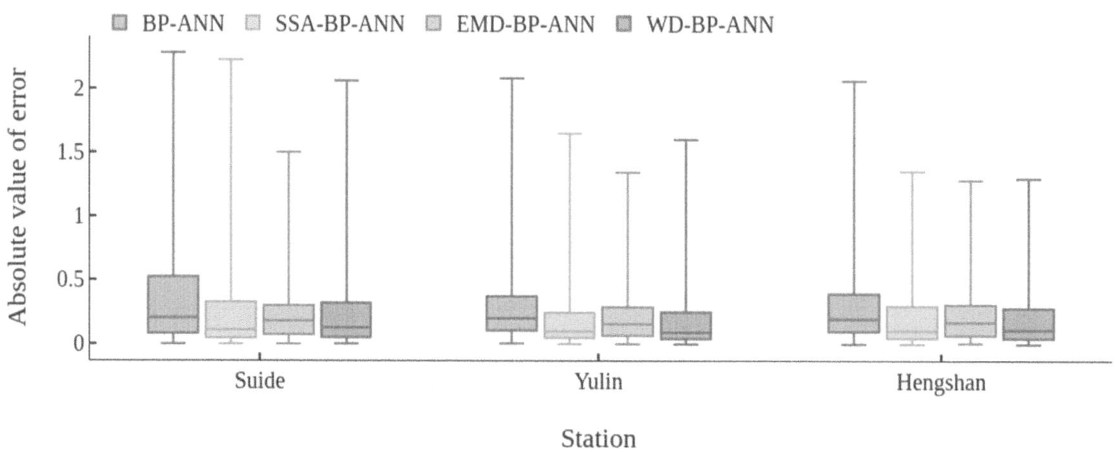

Figure 15 | Boxplot diagram of four models' absolute value of errors for all stations in the testing set.

2. The SSA method can effectively extract the valuable information hidden in the SPI-12 time series, gather and reconstruct the most predictable components into several meaningful subsequences, thereby reducing the influence of noise.
3. The SSA-BP-ANN model is a promising new approach to predict the SPI in arid regions, such as northern Shaanxi, China.

The current research focused on the singular spectrum decomposition preprocessing of the data, which can significantly improve the performance of the BP-ANN prediction model. In future research, other data preprocessing methods will be used to screen out the best hybrid model.

ACKNOWLEDGEMENTS

This study was funded by the Natural Science Basic Research Program of Shaanxi Province (Grant Nos 2019JLZ-16 and 2019JLZ-15) and the Science and Technology Program of Shaanxi Province (Grant Nos 2019slkj-13, 2020slkj-16 and 2018slkj-4). The authors thank the editor for their comments and suggestions.

CONFLICT OF INTEREST

The authors have no relevant financial or non-financial interests to disclose.

DATA AVAILABILITY STATEMENT

All relevant data are included in the paper or its Supplementary Information.

REFERENCES

Achour, K., Meddi, M., Zeroual, A., Bouabdelli, S., Maccioni, P. & Moramarco, T. 2020 Spatio-temporal analysis and forecasting of drought in the plains of northwestern Algeria using the standardized precipitation index. *Journal of Earth System Science* **129** (1), 1–22. doi:10.1007/s12040-019-1306-3.

Anshuka, A., van Ogtrop, F. F. & Vervoort, R. W. 2019 Drought forecasting through statistical models using standardised precipitation index: a systematic review and meta-regression analysis. *Natural Hazards* **97** (2), 955–977. doi:10.1007/s11069-019-03665-6.

Balti, H., Abbes, A. B., Mellouli, N., Farah, I. R., Sang, Y. & Lamolle, M. 2020 A review of drought monitoring with big data: issues, methods, challenges and research directions. *Ecological Informatics* **60**, 101136. doi:10.1016/j.ecoinf.2020.101136.

Başakın, E. E., Ekmekcioğlu, Ö., Çıtakoğlu, H. & Özger, M. 2021 A new insight to the wind speed forecasting: robust multi-stage ensemble soft computing approach based on pre-processing uncertainty assessment. *Neural Computing and Applications*, 1–30. doi:10.1007/s00521-021-06424-6.

Baydaroğlu, Ö., Koçak, K. & Duran, K. 2017 River flow prediction using hybrid models of support vector regression with the wavelet transform, singular spectrum analysis and chaotic approach. *Meteorology and Atmospheric Physics* **130** (3), 349–359. doi:10.1007/s00703-017-0518-9.

Bazrafshan, J., Hejabi, S. & Rahimi, J. 2014 Drought monitoring using the multivariate standardized precipitation index (MSPI). *Water Resources Management* **28** (4), 1045–1060. doi:10.1007/s11269-014-0533-2.

Beguería, S., Vicente-Serrano, S. M., Reig, F. & Latorre, B. 2014 Standardized precipitation evapotranspiration index (SPEI) revisited: parameter fitting, evapotranspiration models, tools, datasets and drought monitoring. *International Journal of Climatology* **34** (10), 3001–3023. doi:10.1002/joc.3887.

Belayneh, A., Adamowski, J., Khalil, B. & Ozga-Zielinski, B. 2014 Long-term SPI drought forecasting in the Awash River Basin in Ethiopia using wavelet neural network and wavelet support vector regression models. *Journal of Hydrology* **508**, 418–429. doi:10.1016/j.jhydrol.2013.10.052.

Belayneh, A., Adamowski, J., Khalil, B. & Quilty, J. 2016 Coupling machine learning methods with wavelet transforms and the bootstrap and boosting ensemble approaches for drought prediction. *Atmospheric Research* **172**, 37–47. doi:10.1016/j.atmosres.2015.12.017.

Bonizzi, P., Karel, J. M., Meste, O. & Peeters, R. L. 2014 Singular spectrum decomposition: a new method for time series decomposition. *Advances in Adaptive Data Analysis* **6** (04), 1450011. doi:10.1142/s1793536914500113.

Bouaziz, M., Medhioub, E. & Csaplovisc, E. 2021 A machine learning model for drought tracking and forecasting using remote precipitation data and a standardized precipitation index from arid regions. *Journal of Arid Environments* **189**, 104478. doi:10.1016/j.jaridenv.2021.104478.

Chen, S., Zhang, L., Zhang, Y., Guo, M. & Liu, X. 2020 Evaluation of Tropical Rainfall Measuring Mission (TRMM) satellite precipitation products for drought monitoring over the middle and lower reaches of the Yangtze River Basin, China. *Journal of Geographical Sciences* **30** (1), 53–67. doi:10.1007/s11442-020-1714-y.

Dai, A. 2011 Drought under global warming: a review. *Wiley Interdisciplinary Reviews: Climate Change* **2** (1), 45–65. doi:10.1002/wcc.81.

Dawson, C. W., Abrahart, R. J. & See, L. M. 2007 Hydrotest: a web-based toolbox of evaluation metrics for the standardised assessment of hydrological forecasts. *Environmental Modelling & Software* **22** (7), 1034–1052. doi:10.1016/j.envsoft.2006.06.008.

Djerbouai, S. & Souag-Gamane, D. 2016 Drought forecasting using neural networks, wavelet neural networks, and stochastic models: case of the Algerois Basin in North Algeria. *Water Resources Management* **30** (7), 2445–2464. doi:10.1007/s11269-016-1298-6.

Durdu, Ö. F. 2010 Application of linear stochastic models for drought forecasting in the Büyük Menderes river basin, western Turkey. *Stochastic Environmental Research and Risk Assessment* **24** (8), 1145–1162. doi:10.1007/s00477-010-0366-3.

Fung, K. F., Huang, Y. F., Koo, C. H. & Mirzaei, M. 2019 Improved SVR machine learning models for agricultural drought prediction at downstream of Langat River Basin, Malaysia. *Journal of Water and Climate Change* **11** (4), 1383–1398. doi:10.2166/wcc.2019.295.

Hagan, M. T. & Menhaj, M. B. 1994 Training feedforward networks with the Marquardt algorithm. *IEEE Transactions on Neural Networks* **5** (6), 989–993. doi:10.1109/72.329697.

Haile, G. G., Tang, Q., Li, W., Liu, X. & Zhang, X. 2019 Drought: progress in broadening its understanding. *Wiley Interdisciplinary Reviews: Water* **7** (2), e1407. doi:10.1002/wat2.1407.

Hassanzadeh, Y., Ghazvinian, M., Abdi, A., Baharvand, S. & Jozaghi, A. 2020 Prediction of short and long-term droughts using artificial neural networks and hydro-meteorological variables. arXiv preprint arXiv:2006.02581.

Hayes, M., Svoboda, M., Wall, N. & Widhalm, M. 2011 The Lincoln declaration on drought indices: universal meteorological drought index recommended. *Bulletin of the American Meteorological Society* **92** (4), 485–488. doi:10.1175/2010BAMS3103.1.

Hinge, G. & Sharma, A. 2021 Comparison of wavelet and machine learning methods for regional drought prediction. In: *EGU General Assembly Conference Abstracts*, pp. EGU21-218.

Kaur, A. & Sood, S. K. 2020 Deep learning based drought assessment and prediction framework. *Ecological Informatics* **57**, 101067. doi:10.1016/j.ecoinf.2020.101067.

Keyantash, J. 2018 *The Climate Data Guide: Standardized Precipitation Index (SPI)*. National Center for Atmospheric Research Staff (Eds). Available from: https://climatedataguide.ucar.edu/climate-data/standardized-pre-cipitation-index-spi (last modified 7 August 2018).

Khan, M. M. H., Muhammad, N. S. & El-Shafie, A. 2018 Wavelet-ANN versus ANN-based model for hydrometeorological drought forecasting. *Water* **10** (8), 998. doi:10.3390/w10080998.

Khan, M. M. H., Muhammad, N. S. & El-Shafie, A. 2020 Wavelet based hybrid ANN-ARIMA models for meteorological drought forecasting. *Journal of Hydrology* **590**, 125380. doi:10.1016/j.jhydrol.2020.125380.

Kimwatu, D. M., Mundia, C. N. & Makokha, G. O. 2021 Developing a new socio-economic drought index for monitoring drought proliferation: a case study of Upper Ewaso Ngiro River Basin in Kenya. *Environmental Monitoring and Assessment* **193** (4), 1–22. doi:10.1007/s10661-021-08989-0.

Kisi, O., Gorgij, A. D., Zounemat-Kermani, M., Mahdavi-Meymand, A. & Kim, S. 2019 Drought forecasting using novel heuristic methods in a semi-arid environment. *Journal of Hydrology* **578**, 124053. doi:10.1016/j.jhydrol.2019.124053.

Kolachian, R. & Saghafian, B. 2021 Hydrological drought class early warning using support vector machines and rough sets. *Environmental Earth Sciences* **80** (11), 1–15. doi:10.1007/s12665-021-09536-3.

Kousari, M. R., Hosseini, M. E., Ahani, H. & Hakimelahi, H. 2017 Introducing an operational method to forecast long-term regional drought based on the application of artificial intelligence capabilities. *Theoretical and Applied Climatology* **127** (1–2), 361–380. doi:10.1007/s00704-015-1624-6.

Liu, X., Zhu, X., Zhang, Q., Yang, T., Pan, Y. & Sun, P. 2020 A remote sensing and artificial neural network-based integrated agricultural drought index: index development and applications. *Catena* **186**, 104394. doi:10.1016/j.catena.2019.104394.

Lloyd-Hughes, B. & Saunders, M. A. 2002 A drought climatology for Europe. *International Journal of Climatology: A Journal of the Royal Meteorological Society* **22** (13), 1571–1592. doi:10.1002/joc.846.

McKee, T. B., Doesken, N. J. & Kleist, J. 1993 The relationship of drought frequency and duration to time scales. In: *Proceedings of the 8th Conference on Applied Climatology*, pp. 179–183.

Mehdizadeh, S., Ahmadi, F., Mehr, A. D. & Safari, M. J. S. 2020 Drought modeling using classic time series and hybrid wavelet-gene expression programming models. *Journal of Hydrology* **587**, 125017. doi:10.1016/j.jhydrol.2020.125017.

Mi, X., Liu, H. & Li, Y. 2019 Wind speed prediction model using singular spectrum analysis, empirical mode decomposition and convolutional support vector machine. *Energy Conversion and Management* **180**, 196–205. doi:10.1016/j.enconman.2018.11.006.

Mishra, A. K. & Desai, V. R. 2006 Drought forecasting using feed-forward recursive neural network. *Ecological Modelling* **198** (1–2), 127–138. doi:10.1016/j.ecolmodel.2006.04.017.

Mokhtarzad, M., Eskandari, F., Vanjani, N. J. & Arabasadi, A. 2017 Drought forecasting by ANN, ANFIS, and SVM and comparison of the models. *Environmental Earth Sciences* **76** (21), 1–10. doi:10.1007/s12665-017-7064-0.

Moreira, E., Russo, A. & Trigo, R. M. 2018 Monthly prediction of drought classes using log-linear models under the influence of NAO for early-warning of drought and water management. *Water* **10** (1), 65. doi:10.3390/w10010065.

Mouatadid, S., Raj, N., Deo, R. C. & Adamowski, J. F. 2018 Input selection and data-driven model performance optimization to predict the Standardized Precipitation and Evaporation Index in a drought-prone region. *Atmospheric Research* **212**, 130–149. doi:10.1016/j.atmosres.2018.05.012.

Özger, M., Başakın, E. E., Ekmekcioğlu, Ö. & Hacısüleyman, V. 2020 Comparison of wavelet and empirical mode decomposition hybrid models in drought prediction. *Computers and Electronics in Agriculture* **179**, 105851. doi:10.1016/j.compag.2020.105851.

Palmer, W. C. 1965 *Meteorological Drought.* U.S. Weather Bureau, Research Paper No. 45, p. 58. Available from: https://www.ncdc.noaa.gov/tempand-precip/drought/docs/palmer.pdf (accessed February 2022).

Pham, Q. B., Yang, T. C., Kuo, C. M., Tseng, H. W. & Yu, P. S. 2021 Coupling singular spectrum analysis with least square support vector machine to improve accuracy of SPI drought forecasting. *Water Resources Management* **35** (3), 847–868. doi:10.1007/s11269-020-02746-7.

Qaisrani, Z. N., Nuthammachot, N. & Techato, K. 2021 Drought monitoring based on Standardized Precipitation Index and Standardized Precipitation Evapotranspiration Index in the arid zone of Balochistan province, Pakistan. *Arabian Journal of Geosciences* **14** (1), 1–13. doi:10.1007/s12517-020-06302-w.

Rehana, S. & Naidu, G. S. 2021 Development of hydro-meteorological drought index under climate change–semi-arid river basin of Peninsular India. *Journal of Hydrology* **594**, 125973. doi:10.1016/j.jhydrol.2021.125973.

Rumelhart, D. E., Hinton, G. E. & Williams, R. J. 1986 Learning representations by back-propagating errors. *Nature* **323** (6088), 533–536. doi:10.1038/323533a0.

Soh, Y. W., Koo, C. H., Huang, Y. F. & Fung, K. F. 2018 Application of artificial intelligence models for the prediction of standardized precipitation evapotranspiration index (SPEI) at Langat River Basin, Malaysia. *Computers and Electronics in Agriculture* **144**, 164–173. doi:10.1016/j.compag.2017.12.002.

Tahroudi, M. N., Ramezani, Y., De Michele, C. & Mirabbasi, R. 2020 A new method for joint frequency analysis of modified precipitation anomaly percentage and streamflow drought index based on the conditional density of copula functions. *Water Resources Management* **34** (13), 4217–4231. doi:10.1007/s11269-020-02666-6.

Thom, H. C. 1958 A note on the gamma distribution. *Monthly Weather Review* **86** (4), 117–122. doi:10.1175/1520-0493(1958)086 < 0117:ANOTGD > 2.0.CO;2.

Vidyarthi, V. K. & Jain, A. 2020 Knowledge extraction from trained ANN drought classification model. *Journal of Hydrology* **585**, 124804. doi:10.1016/j.jhydrol.2020.124804.

Wang, X., Zhang, Y., Feng, X., Feng, Y., Xue, Y. & Pan, N. 2017 Analysis and application of drought characteristics based on run theory and Copula function. *Transactions of the Chinese Society of Agricultural Engineering* **33** (10), 206–214. doi:10.11975/j.issn.1002-6819.2017.10.027.

Wang, X., Zhuo, L., Li, C., Engel, B. A., Sun, S. & Wang, Y. 2020 Temporal and spatial evolution trends of drought in northern Shaanxi of China: 1960–2100. *Theoretical and Applied Climatology* **139** (3), 965–979. doi:10.1007/s00704-019-03024-2.

Waseem, M., Ajmal, M. & Kim, T. W. 2015 Development of a new composite drought index for multivariate drought assessment. *Journal of Hydrology* **527**, 30–37. doi:10.1016/j.jhydrol.2015.04.044.

Wei, S., Zuo, D. & Song, J. 2012 Improving prediction accuracy of river discharge time series using a Wavelet-NAR artificial neural network. *Journal of Hydroinformatics* **14** (4), 974–991. doi:10.2166/hydro.2012.143.

Wu, C., Yeh, P. J. F., Chen, Y. Y., Lv, W., Hu, B. X. & Huang, G. 2021 Copula-based risk evaluation of global meteorological drought in the 21st century based on CMIP5 multi-model ensemble projections. *Journal of Hydrology* **598**, 126265. doi:10.1016/j.jhydrol.2021.126265.

Yu, C., Li, Y. & Zhang, M. 2017 An improved wavelet transform using singular spectrum analysis for wind speed forecasting based on Elman neural network. *Energy Conversion and Management* **148**, 895–904. doi:10.1016/j.enconman.2017.05.063.

Zhang, B., Zhao, X., Jin, J. & Wu, P. 2015a Development and evaluation of a physically based multiscalar drought index: the Standardized Moisture Anomaly Index. *Journal of Geophysical Research: Atmospheres* **120** (22), 11–575. doi:10.1002/2015jd023772.

Zhang, Q., Yao, Y., Li, Y., Luo, Z., Zhang, C., Li, D., Wang, R., Wang, J., Chen, T., Xiao, G., Zhang, S., Wang, S., Guo, N., Bai, H., Xie, J., Yang, X., Dong, A., Deng, Z., Ke, X. & Xu, G. 2015b Research progress and prospect on the monitoring and early warning and mitigation technology of meteorological drought disaster in northwest China. *Advances in Earth Science* **30** (2), 196–211. doi:10.11867/j.issn.1001-8166.2015.02.0196.

Zhang, Y., Li, W., Chen, Q., Pu, X. & Xiang, L. 2017 Multi-models for SPI drought forecasting in the north of Haihe River Basin, China. *Stochastic Environmental Research and Risk Assessment* **31** (10), 2471–2481. doi:10.1007/s00477-017-1437-5.

Zhang, G., Ali, S., Wang, X., Wang, G., Pan, Z. & Zhang, J. 2019a SPI-based drought simulation and prediction using ARMA-GARCH model. *Applied Mathematics and Computation* **355**, 96–107. doi:10.1016/j.amc.2019.02.058.

Zhang, R., Chen, Z. Y., Xu, L. J. & Ou, C. Q. 2019b Meteorological drought forecasting based on a statistical model with machine learning techniques in Shaanxi province, China. *Science of the Total Environment* **665**, 338–346. doi:10.1016/j.scitotenv.2019.01.431.

Zhang, Y., Hao, Z., Feng, S., Zhang, X., Xu, Y. & Hao, F. 2021 Agricultural drought prediction in China based on drought propagation and large-scale drivers. *Agricultural Water Management* **255**, 107028. doi:10.1016/j.agwat.2021.107028.

Zhao, S., Chen, J., Wang, H. & Jin, Q. 2010 Research on drought disaster assessment based on information diffusion and BP neural network. In *2010 International Conference on E-Product E-Service and E-Entertainment*, pp. 1–4. doi:10.1109/iceee.2010.5660314.

Zuo, G., Luo, J., Wang, N., Lian, Y. & He, X. 2020 Two-stage variational mode decomposition and support vector regression for streamflow forecasting. *Hydrology and Earth System Sciences* **24** (11), 5491–5518. doi:10.5194/hess-24-5491-2020.

First received 1 November 2021; accepted in revised form 6 February 2022. Available online 18 February 2022

IWA Publishing's authorised EU representative for General Product Safety Regulations is Diane D'Arras, 15 rue Duret, 75116 Paris, France, e-mail: safety@iwap.co.uk.

Printed and bound by CPI Group (UK) Ltd, Croydon, CR0 4YY
16/06/2026
02140362-0001